The Golden Age of Philosophy
of Science

Also available from Bloomsbury

The Bloomsbury Companion to the Philosophy of Science, edited by
Steven French and Juha Saatsi
Getting Science Wrong, by Paul Dicken
God, Physics, and the Philosophy of Science, by Jeffrey Koperski
The History and Philosophy of Science: A Reader, edited by Daniel J. McKaughan
and Holly VandeWall

The Golden Age of Philosophy of Science 1945 to 2000

Logical Reconstructionism, Descriptivism, Normative Naturalism, and Foundationalism

John Losee

BLOOMSBURY ACADEMIC
LONDON · NEW YORK · OXFORD · NEW DELHI · SYDNEY

BLOOMSBURY ACADEMIC
Bloomsbury Publishing Plc
50 Bedford Square, London, WC1B 3DP, UK
1385 Broadway, New York, NY 10018, USA

BLOOMSBURY, BLOOMSBURY ACADEMIC and the Diana logo are
trademarks of Bloomsbury Publishing Plc

First published in Great Britain 2019
Paperback edition published 2020

Cover design: Eleanor Rose
Cover image © Getty Images

A catalogue record for this book is available from the British Library.

A catalog record for this book is available from the Library of Congress.

ISBN: HB: 978-1-3500-7151-3
       PB: 978-1-3501-6913-5
     ePDF: 978-1-3500-7152-0
    eBook: 978-1-3500-7153-7

Typeset by RefineCatch Limited, Bungay, Suffolk

To find out more about our authors and books visit www.bloomsbury.com
and sign up for our newsletters.

Contents

Introduction

Philosophers take a keen interest in reflexive questions about their discipline. "What is philosophy?" is a perennial topic for debate. Philosophers of science, like philosophers generally, have engaged in debates about the nature and scope of their discipline.

To claim that there exist competing philosophies of science is to claim that there is a discipline which individuals interpret differently. There is general agreement that philosophy of science is a second-order commentary on science, a commentary sensitive to questions about evidential support, theory-choice, and explanatory success.

During the period 1945 to 2000, four positions on the nature and scope of philosophy of science were proposed. These positions were (1) logical reconstructionism, (2) descriptivism, (3) normative naturalism, and (4) foundationalism. One aim of the present study is to formulate a taxonomy applicable to competing philosophies of science. A second aim is to seek answers to four questions that arise from this competition:

1. Should philosophy of science be a normative-prescriptive discipline?
2. If so, can philosophy of science achieve normative-prescriptive status without designating as inviolable some evaluative principles?
3. How ought competing philosophies of science be evaluated? and
4. What is the role of the history of science in this appraisal?

Overview

The Golden Age of Philosophy of Science is a contribution to the history of the philosophy of science. It sets forth the post-Second World War program for the logical analysis of science, the criticisms that displayed the limitations of this approach, and three methodologies proposed to replace it.

Logical reconstructionists dealt with methodological problems and evaluative problems by reference to logical relations among discrete levels of scientific language. Values of scientific concepts were related *via* operational definitions to statements of primary experimental data. These values were explained by formulating deductive (or inductive) arguments with laws as premises. Laws, in turn, were confirmed by logical relations to values of concepts, and ultimately to statements of primary experimental data. Theories were held to explain laws by exhibiting them as logical consequences. Theory-replacement was taken to be progressive if the successor theory incorporated, and extended, the range of its predecessor. A goal of the logical reconstructionist program was to provide explications for epistemological concepts such as "explanation," "confirmation," and "theory-replacement."

Extensive and repeated criticisms of the logical reconstructionist program by Feigl, Achinstein, Feyerabend, and others (1965–1980) had a cumulative impact. Many philosophers of science came to believe that the disparity between actual scientific practice and the orthodox analyses of explanation, confirmation, and theory-replacement was too great.

One response to these criticisms of logical reconstructionism was to reconsider the very nature of the philosophy of science. Some philosophers of science maintained that the proper role of the discipline is the description of actual scientific evaluative practice.

There is a modest version and a robust version of descriptive philosophy of science. The aim of the modest version is the description of the evaluative practice of scientists. The robust version of descriptivism derives from, or superimposes upon, scientific practice a theory about this practice. The theory typically highlights a pattern displayed by, or a set of principles that informs, scientific evaluative practice. Some theories that have been proposed appeal to aspects of the theory of organic evolution. A robust descriptive philosophy of science is a contribution to our understanding of science. It purports to show why science is as it is.

Normative naturalism was a second response to the demise of logical reconstructionism. Normative naturalists deny the logical reconstructionist claim that there exist trans-historical, inviolable evaluative standards. They hold that evaluative standards arise within the practice of science and are subject to revision or rejection in the light of further experience. They claim, nevertheless, that evaluative standards have genuine normative-prescriptive force.

Foundationalists, by contrast, accepted the logical reconstructionist claim that there exist trans-historical inviolable principles that provide support for other statements within science. Candidates for foundational status include intra-theoretical consistency, direct observation reports, experimental laws, evaluative standards and cognitive aims. *The Golden Age of Philosophy of Science* concludes with an examination of the strengths and weaknesses of these alternatives to logical reconstructionism.

Logical Reconstructionism

Sources of Logical Reconstructionism

Philosophy of science is a commentary on science. To whom is this commentary addressed? Norman Campbell, in *Foundations of Science* (1919),[1] announced that he was writing for professional scientists. He proposed to do for physics what Hilbert and Peano had done for mathematics. These mathematicians had clarified the structure of axiom systems, thereby creating a strong foundation for their discipline. Campbell recommended a similar study of the "foundations" of empirical science. He contributed analyses of the nature of measurement, the role of induction in the discovery of scientific laws, and the structure of scientific theories.

Campbell was not the first to pose questions about the structure of science.[2] However, his statement of the aims of philosophy of science was widely shared. After the Second World War, the philosophy of science emerged as a distinct academic discipline, complete with graduate programs and a periodical literature. This professionalization occurred, in part, because many scholars adopted Campbell's outlook. They believed that there were achievements to be won by the study of the foundations of science, and that science would benefit from this study.

In the early 1950s, however, doubts about this program arose. Skeptics suggested that the philosopher is not qualified to prescribe "proper scientific procedure" to the scientist. But then, what task remains? What is it that distinguishes the work of the philosopher of science from that of the historian of science or the sociologist of science? Perhaps the drama includes no role for the philosopher of science.

Five decades after Campbell's study of the foundations of science, Paul Feyerabend cast a retrospective glance over the recent record of the philosophy of science. His conclusion was that

> there is not a single discovery in this field (assuming that there have been discoveries) that would enable us to attack important scientific problems in

a new way or to better understand the manner in which progress was made in the past.[3]

Is the philosophy of science guilty as charged? One purpose of the present study is to compile evidence upon which a verdict may be rendered.

The Legacy of Logical Positivism

To a great extent, the philosophy of science of the 1940s was a legacy of the logical positivism of the 1930s. Rudolf Carnap and Moritz Schlick were the acknowledged leaders of logical positivism.[4] The logical positivist position was that all genuine philosophical problems can be solved by a logical analysis of language, and that metaphysical claims can be shown to fall outside the range of cognitively significant discourse.

The same conviction prevailed in the philosophy of science of the 1940s. Most philosophers of science agreed that an anti-metaphysical bias had promoted the growth of science and that claims about "vital forces," "absolutely simultaneous events," and the "luminiferous aether" had been excluded from science by the same sort of logical analysis recommended by logical positivists.

Logical positivists sought to provide a secure epistemological foundation for the sciences. By and large, they agreed that this was a matter of specifying the ways in which the statements of the sciences are related to "elementary propositions" whose truth or falsity can be determined directly and unambiguously.

Ludwig Wittgenstein outlined the logical structure of this transition in *Tractatus Logico-Philosophicus* (1921).[5] Given a set of logically independent elementary propositions, Wittgenstein constructed an artificial language that contains both complex propositions and universal propositions. Within this language, each complex proposition is a truth-function of its constituent elementary propositions. For example, "p & q" is true provided that p and q both are true, and false otherwise; "p v q" is true provided that at least one of the pair $[p, q]$ is true, and false if both p and q are false. Wittgenstein also specified a procedure to make universally quantified propositions within the language truth-functions of the elementary propositions.

Philosophers of science found this formal structure intriguing. Is it possible to give empirical content to the structure so that the propositions of science may be included and the non-empirical claims of metaphysics may be excluded?

One first would have to select a set of elementary propositions. Wittgenstein had failed to provide a single example of an elementary proposition. However, he did maintain that an elementary proposition (1) exhibits the logical form of a state of affairs, and (2) asserts that this state of affairs exists.[6] It would seem that to determine the truth-status of an elementary proposition one need only ascertain whether the state of affairs asserted to exist does in fact exist.

Philosophers of science who shared Campbell's interest in providing a firm foundation for science were enamored of the prospect of a truth-functional language from which all non-empirical claims are excluded. There were two options for the interpretation of elementary propositions. The first option was the subjectivist position that they are introspective reports of states of immediate conscious awareness. Examples include "Here, now, red spot," "Here, now, click," and "Here, now, bitter taste." The second option was the objectivist position that elementary propositions are intersubjectively observable properties and relations of physical objects. Examples include "This surface is red" and "Object *b* is between *a* and *c*."

In "Testability and Meaning" (1936), Carnap selected the objective version of elementary propositions.[7] Given an extensive set of observation terms that are predicable of individuals, Carnap sought to show how the concepts of the sciences could be generated from this set.

He noted that the concept "arthropod" may be constructed by an explicit definition in terms of phrases reducible to elementary propositions., viz., "*x* is an arthropod" if, and only if, "*x* is an animal and *x* has a segmented body and *x* has jointed legs."[8] Other scientific concepts introduced by explicit definition include "velocity" (v = d/t) and "enthalpy" (H = E + P V).

However, dispositional concepts such as "soluble" and "magnetic field intensity" cannot be defined explicitly in terms of elementary propositions. Given "Sx" = "*x* is soluble," "Px" = "*x* is placed in a liquid," and "Dx" = "*x* dissolves," the explicit definition

$$(x) \ [Sx \equiv (Px \supset Dx)]$$

would qualify as "soluble" every object not now placed in a liquid. If "Pa" is false, then "(Pa $\supset$ Da)" is true, and "Sa" is true, given the meanings of "$\supset$" and "$\equiv$").

Carnap proposed instead conditional, or "contextual," definitions of dispositional concepts.[9] In the case of "soluble" the contextual definition is

$$(x) \ [Px \supset (Sx \equiv Dx)]$$

This contextual definition restricts the assignment of truth values to the phrase "*a* is soluble" to instances in which "*a* is placed in a liquid" is true. Thus the dispositional term "soluble" is given only a partial meaning in terms of elementary propositions.

A further problem for the construction of an empiricist language for science is that certain theoretical concepts cannot be introduced into this language by contextual definitions. Carnap initially maintained (1938) that theoretical concepts could be defined contextually in terms of observation statements.[10] He withdrew this claim in the essay "The Methodological Character of Theoretical Concepts" (1956).[11] He recognized that concepts such as "the velocity of an individual molecule" in the kinetic theory of gases and "the Ψ-function" in quantum mechanics are terms in axiom systems. These concepts cannot be related directly by contextual definitions to elementary propositions. It is the entire axiom system that is related to statements about observables.

"v_i," the velocity of molecule i, is related to the root-mean-square velocity of all the molecules of a gas by the relation

$$\mu = \sqrt{v_1^2 + v_2^2 + v_3^2 + \ldots v_n^{/n}}$$

"μ," in turn, is related to observables "pressure" and "temperature" by the relations "$P = n\,m\,\mu^2/V$" and "$T = n\,m\,\mu^2/2\,k$," where n is the number of molecules of mass m, V is the volume of the gas, and k is Boltzmann's constant.

The Ψ function is "$\Psi = A\,e^{2\pi i h (P x = E t)}$," where A is a constant, e is the base of the system of natural logarithms, $i = \sqrt{-1}$, h is Planck's constant, P is a momentum operator, and E is an energy operator. Since Ψ involves the imaginary number i, it cannot be defined contextually. Since no contextual definitions are available for certain terms in the axiom systems of scientific theories, the project to construct the concepts of science from a basic empirical vocabulary is a failure. The language of science is not an empiricist language.

Had it been possible to provide contextual definitions for theoretical concepts, statements about these concepts would be reducible to observation reports, given the execution of specified operations. The language of science then would be homogeneous. There would be no reason to set apart a distinct level of theoretical language. Logical reconstructionists concluded that, since theoretical concepts cannot be linked to observation statements in this way, there is a two-tier structure within the language of science.

Quantum Physics and the Philosophy of Science

The success of quantum mechanics provided evidence for the two-tier view of scientific language. Many philosophers of science believed that there were epistemological lessons to be drawn from analysis of the structure of quantum theory.

Applications of quantum theory involve transitions from a language in which observations are recorded into a theoretical language, and back again. A quantum-mechanical interpretation of an experiment typically is a three-stage process: (1) translation of an initial experimental situation into a probability function; (2) mathematical analysis of the variation of this function with time; and (3) translation of the results of calculation into a prediction of the results of a new measurement to be made of the system.[12]

Given an "observational language" and a "theoretical language," a quantum-mechanical interpretation may be represented as in Figure 1.

To interpret the situation in this way would seem to presuppose that (1) observational and theoretical languages can be distinguished; (2) the results of experiments can be expressed adequately without reference to the theoretical language of quantum theory (Bohr and Heisenberg insisted that descriptions of experimental arrangements must be given in the language of classical physics[13]); and (3) there are rules for translating statements of the observational language into the theoretical language, and conversely.

Proponents of the Copenhagen interpretation of quantum mechanics employed a semantic rule suggested by Max Born. Born suggested that, if the proper function corresponding to the state of a subatomic particle is Ψ, then $|\Psi|^2$ dV is the probability that the particle is in the volume element dV.[14] The

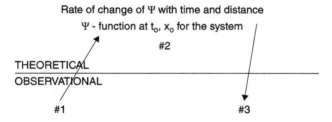

Rate of change of Ψ with time and distance
Ψ - function at t_0, x_0 for the system

#2

THEORETICAL
OBSERVATIONAL

#1   #3

Initial Experimental Situation: **Predicted Result of Further Measurement on the System:**

e.g., Energy and direction of a beam   e.g., Scattering distribution (number
of particles of charge q and mass m of particles as a function of angle of
incident upon a crystal of specified   deflection after passing through the
structure   crystal)

Figure 1 Quantum Theory and Language Levels.

Ψ-function typically has an imaginary component, involving i ($\sqrt{-1}$). Since $i^2 = -1$, $|\Psi|^2$, the square of the absolute magnitude of the amplitude of this function, is a real number that can be correlated with physically significant results of measurement.

The Born interpretation correlates the state of an individual subatomic particle with a probability only. But given an ensemble of particles each in state Ψ_a, the Born interpretation correlates $|\Psi_a|^2$ with observable statistical patterns such as scattering distributions, electron charge densities, and the distributions of electronic states following the excitation of atoms.

Special Relativity Theory and Operationalism

In the late 1920s, P. W. Bridgman sought to derive an "epistemological lesson" from the success of Einstein's theory of special relativity. Einstein had stressed the necessity of links between theoretical claims and observation reports. In so doing, he restated an emphasis made earlier by Mach, Poincare, and Duhem.

Ernst Mach had complained that to define "mass" as "quantity of matter" is a purely verbal exercise.[15] It does not further the purposes of scientific inquiry. He suggested that the term "mass" be introduced by reference to operations performed. This introduction stipulated, in part, that the ratio of two masses is equal to the inverse ratio of the accelerations of the two bodies, measured under specified conditions.

Mach also questioned the usefulness to science of Newton's concepts "absolute space" and "absolute time." There are no operations of measurement that can be performed that yield values of these concepts. Mach recommended that these concepts be eliminated from physics.[16]

Henri Poincare extended Mach's critical approach to the concept of "force." He criticized those who would base an understanding of "force" on a presumed extrapolation from our direct apprehension of effort as we lift or push against heavy objects. According to Poincare, the concept "force" gains importance in science because there are operational procedures by which to measure its values.[17] One such procedure is to measure the extent of deformation of a stretched spring by appeal to Hooke's Law ($F = -k \,\Delta x$).

Pierre Duhem extended the operational requirement from concepts to scientific theories that incorporate them.[18] He maintained that a scientific theory is empirically meaningful only if its applications make assertions about concepts whose values can be measured. Without links to the results of measurement, a putative theory lacks empirical significance.

Duhem was careful to insist, however, that a theory may satisfy the operational requirement even though some of the concepts of the theory are not linked directly to measuring operations. In the kinetic theory of gases, for instance, no operations are specified for individual molecular velocities. The kinetic theory is empirically meaningful, nevertheless, since it makes claims about concepts such as "temperature," "pressure," and "volume," whose values are subject to measurement.

It was Einstein's discussion of the concept of "simultaneity," however, that most impressed Bridgman. Bridgman pointed out that prior to Einstein's work, simultaneity was assumed to be an objective property of two or more events. Events *A* and *B* were believed to have one, and only one, of three possible temporal relations: *A* occurred before *B*, *A* occurred after *B*, or *A* and *B* occurred simultaneously.

Einstein asked how a physicist could establish the simultaneity of two events.[19] He noted that any judgment of simultaneity involves measurements of the events by an observer. Such measurement, in turn, presupposes a transfer of information from the events to the observer by means of a signal. However, since a transfer of information cannot be instantaneous, judgments about the simultaneous occurrence of two events must depend on the relative motions of the systems which contain the events and the observers. Given a particular set of motions, observer Smith on system (1) may judge that event *A* on system (1) and event *B* on system (2) are simultaneous. Observer Jones on system (2) may judge otherwise. There is no reason *a priori* to prefer Smith's determination to that of Jones. Einstein concluded that "simultaneity" is predicated correctly only of a relationship involving two or more events and an observer. The concept "absolute simultaneity" cannot be linked to measuring operations and for that reason is not an empirically meaningful concept.

Bridgman's Early Position

Bridgman was greatly impressed by the success of the theory of special relativity, and he resolved to draw from Einstein's achievement an epistemological lesson. One important conclusion of Einstein was that the concept "absolute simultaneity" has no empirical significance. Bridgman suggested that as a general principle that all concepts which are not linked to measuring procedures should be excluded from physics. He observed that

> if by convention we agree to use only those concepts in describing physical situations to which we can give a meaning in terms of physical operations, then we are sure that we shall not have to retract.[20]

Operational definitions link scientific concepts to statements that record the results of measurements. The form of an operational definition is

$$(x) [Ox \supset (Cx \equiv Rx)]$$

where　Ox = operation O is performed,
　　　　Cx = concept C applies (or has a specified value), and
　　　　Rx = the result of performing the operation is R.

The operational formula provides a partial definition at best. It takes the form of a conditional claim—e.g., *if* a substance is placed in water then it is soluble in water if, and only if, it dissolves. Nothing is said about the substance in contexts in which it is not placed in water. Gustav Bergmann observed that

> the important footnote operationalism has contributed ... [is that] we often must do something, manipulatively, if we want to find out whether a certain statement is true.[21]

Operational definitions provide links between two levels of language within science—the level within which values are assigned to scientific concepts and the level within which basic observational data are recorded. Given an operational definition and statements about a measuring procedure and its result, one can deduce a value for the concept defined. Suppose that

　　　　$Pa = a$ is an instance in which a substance is placed in water, and
　　　　$Da = a$ is an instance in which the substance dissolves;

it follows that a is an instance in which the substance is soluble in water, viz.,

$$(x) [Px \supset (Sx \equiv Dx)]$$

$$\underline{Pa \ \& \ Da }$$

$$\therefore Sa$$

Other concepts introduced by straightforward operational definitions in terms of instrumental procedures include:

"an object is electrically charged"—$(x) [Nx \supset (Ex \equiv Dx)]$

where　$Nx = x$ is a case in which an object is brought near a neutral
　　　　　electroscope,
　　　　$Ex = x$ is a case in which an object is electrically charged, and
　　　　$Dx = x$ is a case in which the leaves of the electroscope diverge;

"an object has local length l"—(x) [Rx ⊃ (Lx ≡ Sx)]

where Rx = a properly calibrated measuring rod is placed on the object,

          Lx = the length of the object is l, and

          Sx = the difference between the points on the scale that correspond to the ends of the object is l;

"an object has temperature t"—(x) [Px ⊃ (Tx ≡ Mx)]

where Px = a thermometer is placed in (or on) an object,

          Tx = the temperature of the object is t, and

          Mx = the meniscus of the thermometer is on the t-line.

Some concepts are associated with more than one operational definition. The above definition introduces "temperature" by reference to operations performed with a thermometer. Other operational definitions of "temperature" are available, e.g., by reference to operations performed with a thermocouple.

Various instrumental procedures are available to assign values to "temporal interval," among them operations involving water clocks, spring-driven clocks, pendulum clocks, the rotation of the Earth and the vibrations of cesium atoms. Henry Margenau pointed out that scientists base the selection of a particular operational definition on theoretical considerations.[22] For example, scientists prefer to utilize "time interval determined by reference to the swings of a pendulum" rather than "time interval determined by the weight of water flowing through a hole in a bucket." They do so, in part, because data that confirms a law of nature such as the law of falling bodies show less deviation from calculated values when time is measured by reference to a pendulum clock.

Bridgman acknowledged that the relation between values assigned to a concept and measuring procedures may be complex. He noted, for instance, that the stress within a deformed elastic body is computed *via* mathematical theory from measurements of strain on the surface of the body. He held that "stress" nevertheless is a *bona fide* scientific concept which satisfies the operational criterion of demarcation.[23]

Assigning a value to the concept "entropy change in an irreversible process" (e.g., the Joule-Thomson expansion of a gas into a vacuum) is a complex process. It is calculated by first devising a *reversible* path between the same initial and final states and then measuring the heat absorbed and the temperature at which it is absorbed in the reversible process.

The Ψ-function of quantum theory also is correlated only indirectly with instrumental procedures. Values of Ψ cannot be determined by means of instrumental operations because the function contains the imaginary number $\sqrt{-1}$. Nevertheless, the square of the absolute magnitude of the function—$|\Psi|^2$—is correlated with instrumentally determinable values of scattering distributions, electron charge densities, and orbital transition frequencies.

"Stress," "entropy change in an irreversible process," and the "Ψ-function" satisfy the operational criterion of empirical meaningfulness. Bridgman held that the case is otherwise for "absolute space," "absolute time," and the notion that as the solar system moves through space, both measuring instruments and the dimensions of objects measured contract at the same rate.[24]

In *The Logic of Modern Physics* (1927), Bridgman made a further claim for operational analysis. In addition to recommending an operational criterion of demarcation for scientific concepts, he affirmed an operational theory of meaning. Bridgman claimed that the meaning of a concept is nothing over and above the operations performed to assign values to it. He declared that

> the concept of length is ... fixed when the operations by which length is measured are fixed: that is, the concept of length involves as much as and nothing more than the set of operations by which length is determined. In general, we mean by any concept nothing more than a set of operations: *the concept is synonymous with the corresponding set of operations.*[25]

The concept "temperature," for example, means nothing more than the operations by which values of temperature are assigned. Hence, "temperature measured by a thermometer" and "temperature measured by a thermocouple" must be treated as two distinct concepts.

This extreme claim was widely criticized. Bergmann maintained that Bridgman's identification of meaning and instrumental operations is "unduly restrictive" because "it excludes the terms of interpreted axiomatic calculi."[26] Carl Hempel also rejected Bridgman's claim. He noted that

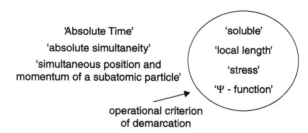

Figure 2 The Operational Criterion of Empirical Significance.

an operational definition of a concept . . . will have to be understood as ascribing the concept to all those cases that *would* exhibit the characteristic response if the test conditions *should* be realized. A concept thus characterized is clearly not "synonymous with the corresponding set of operations." It constitutes not a manifest but a potential character, namely, a disposition to exhibit a certain characteristic response under specified test conditions.[27]

Bridgman's Revised Position

In his later writings, Bridgman reaffirmed the operational criterion of demarcation for scientific concepts. A *bona fide* scientific concept must be linked, however indirectly, to measuring procedures. However, he revised his earlier claim that the meaning of a concept is synonymous with operations performed to apply it in specific situations. His later position was that concepts can be said to be synonymous with operations only if we count as "operations" any conscious activities whatever. In this broad sense of "operation," it would be redundant to speak of "operational definitions."

However, the context in which Bridgman's initial statement of the operational theory of meaning had been given was a discussion of instrumental procedures for assigning values to such concepts as "length" and "time." Some critics assumed that Bridgman's early position was that the meaning of an empirically significant concept is the *physical* operations performed in assigning values to it. R. B. Lindsay is one critic who accused Bridgman of defending this position.[28] However, a careful reading of Bridgman's discussions in *The Nature of Physical Reality* (1936) and *The Logic of Modern Physics* (1960) supports Bridgman's own claim that he never had defended this position. Actually, Bridgman's position on concepts and physical operations was similar to that of Duhem. Bridgman insisted that *some* of the concepts of a theory must be linked to operations of measurement. However, within a theory that satisfies this criterion of demarcation, he allowed concepts defined only by "pencil and paper operations" which link them to other concepts.

In his later writings, Bridgman also discussed certain inherent limitations of operational analysis. One limitation is that it is not possible to specify all the conditions present when an operation is carried out. However, to be of use in science, an operation must be capable of being repeated by any qualified observer. In practice, scientists assume that a particular operation can be repeated independently of the variation of a large number of factors that are not taken into account. Whether this assumption is justified is itself a matter of experience. Bridgman noted that in many cases the assumption does seem to be justified, but he cautioned that an extension of formerly useful operations to new domains of experience may force scientists to take account of conditions previously ignored.

A second limitation of operational analysis is the necessity to accept some unanalyzed operations. For practical reasons, the analysis of operations in terms of other operations cannot proceed indefinitely. For example, the concept "dissolves" may be analyzed in terms of operations which establish uniformity of composition. Such operations may include sampling techniques, and weight and volume determinations. The operation of weighing, in turn, may be analyzed further by specifying the methods for constructing and calibrating balances. But such analysis terminates at some point. Scientists usually accept certain types of operations without calling for further analysis. For example, provided that standard precautions about parallax are observed, scientists accept as unanalyzed reports of pointer readings on the scales of instruments. To be sure, the instrument may not be functioning properly and the reported value may be incorrect, but the determination of just where the pointer is on a scale normally is accepted as an operation needing no further analysis. This is true both in classical physics and quantum physics.

Bridgman observed that our experience to date has been such that no difficulties have arisen from accepting as unanalyzed this type of operation. He conceded, however, that, one day, extension of our experience may suggest a need for further analysis. Bridgman maintained that, although the analysis of operations in terms of other operations must stop somewhere, the decision to accept a particular unanalyzed operation is open to challenge. Moreover, he maintained that one always can respond to such a challenge by giving a more detailed specification of operations.[29]

The essential claim made for operational analysis in Bridgman's later writings is that it is a useful empirical method. Bridgman maintained that theory-construction is facilitated by focusing attention on operations performed in applying concepts in specific physical situations. Scientists who assume the operational point of view are unlikely to hypostatize their concepts and become bogged down in disputes over the "real nature" of the realities connoted by such concepts as "absolute space," "caloric," and "phlogiston." According to Bridgman, the ultimate justification of the operational approach is a pragmatic justification. He declared that

> we have observed after much experience that if we want to do certain kinds of things with our concepts, our concepts had better be constructed in certain ways.[30]

The Orthodox Program for the Philosophy of Science

During the early post-war period, most philosophers of science accepted Hans Reichenbach's distinction between the context of discovery and the context of justification.[31] They agreed that the proper domain of philosophy of science is the context of justification. The central problems of the discipline are problems about explanation and confirmation.

In addition, most philosophers of science were committed to a logical reconstructionist orientation. This orientation was based on certain assumptions about the nature and scope of philosophy of science:

1. The theories, laws, and experimental findings of science are, or may be expressed as, declarative sentences.
2. These sentences may be reconstructed in the symbolism of formal logic.
3. The task of philosophy of science is to specify criteria for the evaluation of laws and theories. Once suitable criteria have been formulated, they may be applied, for instance, to assess the adequacy of proposed explanations, to determine the degree of support provided a law by observational evidence, or to gauge the rationality of instances of theory-replacement.
4. To formulate an evaluative criterion is to specify an "explication" of the appropriate epistemological term, e.g., "law," "theory," "confirmation," "explanation," "reduction," et al. An explication is a logical relation among sentences.
5. The evaluation of laws and theories is possible because there exists a theory-neutral language in which the results of observation are recorded.
6. Questions about the rationality of developments in the history of science can be decided by comparing the theories which were dominant within a domain of science at different times.
7. Although the history of science may provide clues about the relations among theories, laws, and observation reports, what is important to the philosophy of science is the logic of these relations.
8. Proposed explications of an epistemic concept are subject to appraisal at a higher level of analysis. The appraisal requires a comparison with our pre-analytic understanding of the concept. An explication is successful only if it is recognized to be a refinement of our pre-analytic understanding. Applications of the explicated concept must conform, at least roughly, to our pre-analytic intuitions about the concept. For instance, an explication of "*e* confirms *H*" is successful only if its applications to specific evidence statements and hypotheses are recognized to be cases of "confirmation."

However, a successful explication also must improve upon pre-analytic usage. Improvement may be achieved in various ways, among them expansion of the range of application of the pre-analytic concept, resolution of "grey-area" cases of applicability, and demonstration that certain uses of the pre-analytic concept are incorrect.

The logical reconstructionist aim is to prescribe an explicated concept as an improvement upon actual usage. Non-coincidence with actual usage does not, in itself, invalidate an explication. On the other hand, an explication must bear *some* resemblance to our pre-analytic understanding.

In 1945 it seemed evident that the language of science includes an observational level and a theoretical level. The new theories in physics and the studies of logical positivists had provided support for this distinction. This two-tier picture has a fine structure. A hierarchy of language levels may be delineated within the language of science. At the base of the hierarchy are statements that record instrument readings; at the apex are theories (see Figure 3 below).

Post-War orthodox philosophy of science is a second-order commentary on this first-order subject matter. Its aim was to uncover the logical relationships among the various levels. Carnap and Bridgman had shown that there is a deductive relationship between the bottom two levels. Given an operational definition and the appropriate experimental data, one can deduce a statement that assigns a value to a scientific concept, for example, that a particular solution is acidic, viz.:

$$(x) \ [Bx \supset (Ax \equiv Rx)]$$
$$Ba$$
$$Ra$$
$$\overline{}$$
$$\therefore Aa$$

Level	Content	E.g.
Theories	Deductive systems in which laws are theorems	Kinetic Molecular Theory
Laws	Invariant (or statistical) relations among concepts	Boyle's Law ($P = 1/V$)
Values of concepts	Statements that assign values to concepts	'P = 2.0 atm.' 'V = 1.5 lit.'
Primary experimental data	Statements about pointer readings, menisci, counter clicks, *etc.*	'Pointer P is on 3.6'

Figure 3 Language Levels in Science.

where $Bx = x$ is a case in which a piece of blue litmus paper is placed in a liquid,

$Rx = x$ is a case in which the litmus paper turns red, and

$Ax = x$ is a case in which the solution is acidic.

Ba and Ra are statements about operations and the results thereof. Aa is a statement that assigns a value to a scientific concept. These two levels of scientific language are deductively related, viz.:

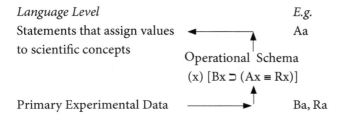

Language Level   *E.g.*
Statements that assign values   Aa
to scientific concepts
                              Operational Schema
                              (x) [Bx ⊃ (Ax ≡ Rx)]

Primary Experimental Data   Ba, Ra

Patterns of Explanation

Deductive-Nomological Explanation

Encouraged by the results achieved by Carnap and Bridgman, orthodox theorists sought to uncover the logic of the relations between the next uppermost levels of scientific language. Scientific laws state invariant or statistical relations among concepts. Laws are invoked to explain why concepts have certain values, and the values of concepts are cited, in turn, as confirming evidence for laws.

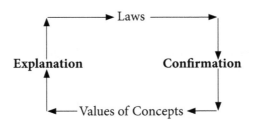

Aristotle expressed the aim of science as a progression from knowledge of facts to knowledge of the reasons why facts are as they are. He maintained that the reason why a fact is as it is is because a set of general principles applies to the case in question. This being the case, one can explain a fact by deducing a statement about it from premises that include appropriate general principles. At the very beginning of philosophical reflection about science, Aristotle formulated a deductive ideal of scientific explanation.

$L_1, L_2, \ldots L_k$ General Laws

$C_1, C_2, \ldots C_r$ Statements of Antecedent Conditions

$\therefore E$ Description of Phenomena

Figure 4 The Deductive-Nomological Pattern of Explanation

The logic of this ideal was set forth in 1948 by Carl Hempel and Paul Oppenheim.[1] They maintained that many scientific explanations conform to the deductive pattern in Figure 4.

An example of this pattern from the history of optics is Newton's explanation of the colors produced when sunlight strikes a prism.

Sunlight comprises rays of different colors.	Laws
Each color is refracted by a prism through a specific and unique angle.	
A beam of sunlight is incident at a certain angle upon a prism of specified geometry.	Antecedent
The prism is at a certain distance from the wall.	Conditions
$\therefore$ A spectrum of colors appears on the wall.	Phenomenon

For many deductive-nomological explanations, antecedent conditions may be subdivided into boundary conditions and initial conditions. Boundary conditions are conditions under which a law is believed to hold. Initial conditions are conditions realized prior to, or at the same time as, the phenomenon to be explained. For example, the expansion of a heated balloon may be explained as follows:

$[(V_1 / V_2) = (T_1 / T_2)]_{,m, P = k}$	Gay-Lussac's Law
Mass and pressure are constant	Boundary conditions
$T_2 = 2\, T_1$	"Initial conditions"
$\therefore V_2 = 2\, V_1$	Phenomenon

Hempel and Oppenheim were not content just to describe a pattern of explanation widely used by scientists. They held that it is a task of the philosophy of science to stipulate necessary and sufficient conditions of "correct" scientific explanations. According to Hempel and Oppenheim, in a correct, or "true," deductive-nomological explanation, the premises

1. imply the conclusion,
2. contain general laws actually used in the derivation,

3. are subject to test by experiment or observation (at least in principle), and
4. are true.

They anticipated that objections would be forthcoming to the fourth of the above requirements. One alternative to requirement (4) might be to insist only that general laws be highly confirmed in the light of the evidence available. Hempel and Oppenheim maintained that this would be to use the concept of explanation in an improper way. We then would have to say that a certain explanation was correct given evidence E, but that it became incorrect when additional evidence E^* had been accumulated. According to Hempel and Oppenheim, the proper interpretation of this situation is that what at one time appeared to be a true law turned out not to be so, and that consequently explanations that utilized it never had been correct.

Hempel and Oppenheim introduced the notion of a "potential explanation" to cover certain arguments whose premises are not known to be true. A "potential explanation" satisfies the first three conditions above, but not the requirement that the premises be true.[2] Just as a correct, or "true," explanation contains at least one law in its premises, a potential explanation contains at least one "lawlike sentence." Hempel conceded that a satisfactory explication of "lawlike sentence" is not available. A lawlike sentence is a sentence which, if true, would count as a law. But of course "what would count as a law" is itself in question.

Eberle, Kaplan, and Montague pointed out in 1961 that the Hempel and Oppenheim conditions for deductive-nomological explanation are too inclusive.[3] There exist arguments that satisfy the Hempel and Oppenheim conditions but fail to explain their conclusions.

On the Eberle, Kaplan, and Montague reconstruction of the Hempel and Oppenheim position, the ordered couple (T, C) is an explanans for singular sentence E, if, and only if,

1. T is a theory,
2. T is not logically equivalent to any singular sentence,
3. C is singular and true,
4. E is derivable from (T, C), and
5. there is a class K of basic sentences such that C is derivable from K, and neither E nor T is derivable from K.

But let e = the Eiffel Tower,
          Cx = "x conducts electricity," and
          Mx = "x is a mermaid."

Eberle, Kaplan, and Montague noted that the following argument counts as a "scientific explanation" on the Hempel and Oppenheim account:

$$T: - \ (x) \ (Mx \supset Cx)$$
$$C: - \ {\sim}Me \supset Ce$$
$$E: - \ \therefore \ Ce$$

Condition (4) is satisfied because the above argument is valid. Condition (5) also is satisfied by selecting K = Me. The argument clearly does not explain why the Eiffel Tower conducts electricity.

Eberle, Kaplan, and Montague developed a number of ways to formulate arguments that both satisfy the Hempel and Oppenheim requirements and fail to explain. In an article in the same issue of *Philosophy of Science*, David Kaplan showed that these types of arguments can be excluded from the class of deductive-nomological explanations by adding three additional requirements to the Hempel and Oppenheim conditions. Kaplan's additional requirements are:

1. if a singular sentence is explainable by a given theory, then it is explainable by any theory from which the given theory is logically derivable
2. any singular sentence which is logically derivable from singular sentences explainable by a theory is itself explainable by that theory; and
3. there is an interpreted language L which contains a fundamental theory T and singular sentences E and E* which are true but not logically provable such that E is explainable by T and E* is not explainable by T.[4]

Hempel accepted the Eberle, Kaplan, and Montague criticism. He emphasized, however, that the addition of Kaplan's requirements does block irrelevant arguments of the Eiffel Tower type.

Inductive-Statistical Explanation

In their paper of 1948, Hempel and Oppenheim stressed that many acceptable scientific explanations are of inductive form.[5] Hempel subsequently indicated that an inductive-statistical pattern is widely exemplified in scientific practice.

p (O, A) = k	Statistical law
i is a case of A	Antecedent conditions
i is a case of O	Description of phenomenon

Figure 5 The Inductive-Statistical Pattern of Explanation

In Figure 5, p (O, A) is the probability that an individual which is an *A* also is an *O*, and *k* is a decimal "close to" 1.0. The double line separating premises and conclusion indicates that the argument does not have deductive force, but that the premises provide only strong inductive support for the conclusion. An example given by Hempel is

A high percentage of patients with streptococcus infections recover within 24 hours after being given penicillin.

Jones had a streptococcus infection and was given penicillin.

Jones recovered from streptococcus infection within 24 hours after being given penicillin.[6]

Other inductive-statistical explanations invoke statistical laws applicable to processes as diverse as radioactive decay, inheritance, and human behavior.

Hempel maintained that scientific explanation may be achieved by either deductive arguments or inductive arguments. In either case the conclusion is subsumed under laws—universal generalizations in the case of deductive-nomological explanations, and statistical generalizations in the case of inductive-statistical explanations.

Hempel and Oppenheim pointed out that there is an important symmetry between deductive explanation and prediction. Both utilize the deductive-nomological pattern. The difference between explanation and prediction is one of intent. In explanation, one begins with the conclusion and creates appropriate premises. In prediction, one begins with laws and statements about antecedent conditions and deduces a statement about a phenomenon to be observed.[7] This alleged symmetry subsequently became the subject of an extended debate (see pp. 115–123).

The Status of Teleological Explanations

Aristotle had maintained that an explanation of a natural process is incomplete unless it specifies the *telos*, the "for what end," of the process. One consequence of the scientific revolution of the seventeenth century was that the concept *telos* was largely expunged from natural philosophy.

In 1943, Arturo Rosenblueth, Norbert Wiener, and Julian Bigelow argued that the concept of "purpose" is of importance in the scientific explanation of behavior.[8] The authors were greatly impressed by recent advances in the design of servomechanisms. Servomechanisms are devices whose output is modified on the basis of input signals received from the environment. An example of this kind of device is the "target-seeking" torpedo whose path is continuously realigned on the basis of sound waves received from the target. Wiener suggested that the natural way to describe such devices is in terms of a capacity "to adjust future conduct by past performance."[9]

It was the intention of Rosenblueth, Wiener, and Bigelow to characterize purposeful activity solely in terms of observable aspects of behavior. Whether or not a particular instance of behavior is purposeful then could be determined by appeal to strictly behavioral criteria.

As a contribution to this program, Rosenblueth, Wiener, and Bigelow proposed the following taxonomy of behavior:

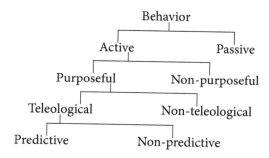

This dichotomous classification scheme is based on the following distinctions:

1. "Active behavior" is behavior in which the behaving object itself is the source of that output energy involved in the behavior. "Passive behavior" is behavior in which the output energy has a source external to the behaving object.

2. "Purposeful active behavior" is behavior which may be interpreted as directed to the attainment of a goal—i.e., to a final condition in which the behaving object reaches a definite correlation in time or in space with respect to another object or event.[10] "Non-purposeful active behavior" is behavior not subject to such an interpretation.

3. "Teleological purposeful behavior" is purposeful behavior in which negative feedback is present. Such behavior is "self-correcting" in that signals from the goal are utilized to modify the behavior of the object. "Non-teleological purposeful behavior" is behavior in which negative feedback is not present.

4. "Predictive teleological behavior" is behavior in which the negative feedback relation involves extrapolation to a future relation between the behaving object and the goal-object or event. The firing of a ground-to-air missile at a

moving airplane would be an example. "Non-predictive teleological behavior" is behavior in which this sort of extrapolation is not present. Rosenblueth, Wiener, and Bigelow suggested as an illustration the pursuit of a mouse by a cat.

The adequacy of the Rosenblueth, Wiener, and Bigelow taxonomy was challenged by Richard Taylor in an article published in 1950.[11] Taylor complained that the characterization given of purposeful behavior by these authors was much too broad. If a sufficient condition of purposeful behavior is that the behaving object reach "a definite correlation in time or in space with respect to another object or event," then a wide variety of processes would qualify as "purposeful." For instance, the return to equilibrium position of a deflected compass needle would qualify as purposeful activity. The rusting of a gun would be as purposeful as the activity of firing it at a target. And the addition of a weight to a roulette wheel so that the wheel invariably stops at "6" would convert its previously "purposeless" behavior to "purposeful" behavior. The Rosenblueth-Wiener-Bigelow criterion of purposeful behavior is far too inclusive.

Taylor pointed out that the Rosenblueth, Wiener, and Bigelow account is deficient in yet a more important respect. Explanations of behavior that invoke the notion of "purpose" are redundant. They add nothing of scientific value to those explanations that cite the relevant causes of the behavior. For example, the motion of a target-seeking torpedo can be explained by appeal to an appropriate theory of sound propagation, an appropriate theory about the steering mechanism, and so forth.

Rosenblueth and Wiener replied to Taylor's criticisms.[12] They rejected the charge that their criterion of purposeful behavior was too inclusive. They conceded that Taylor's characterization of the behavior of compass needles and weighted roulette wheels as "purposeful" was correct. However, they complained that Taylor failed to point out that such behavior is passive rather than active. Rosenblueth and Wiener suggested that passive behavior, as well as active behavior, is subject to the "purposeful–purposeless" dichotomy.

Rosenblueth and Wiener continued to maintain that the concept of purpose is of value in the scientific explanation of behavior. They listed a set of criteria to distinguish purposeful from non-purposeful behavior. An object acts purposefully if

1. it is coupled with other objects in such a way that changes in these objects affect its behavior;

2. it registers messages from its surroundings; and
3. it is "oriented toward or guided by a goal."[13]

Whether these conditions are met in a particular instance is difficult to ascertain. Rosenblueth and Wiener conceded that observers may differ in their evaluation of behavior. They suggested that recognition of purposeful behavior requires observation of the behavior under diverse conditions. Only then can goal-directed behavior be reliably distinguished from random coincidences.[14] However, they insisted that many instances of behavior do qualify as purposive. The observer at a greyhound racetrack, for example, may come to realize that the behavior of the mechanical hare differs from that of the hounds in important respects. One can describe the behavior of the hare without taking into account the behavior of the hounds, but not conversely. Moreover, if the hare is detracked, it displays no tendency to return to its original path. By contrast, if the hounds meet an obstacle, they detour and resume pursuit of the hare. Situations of this kind are common, according to Rosenblueth and Wiener. They emphasized that

> the vast majority of engineers working on the theory or the design of servomechanisms share our conviction that the notion of purpose is essential for their work.[15]

In a rejoinder to the Rosenblueth and Wiener article, Taylor professed to remaining unconvinced that criteria of purposefulness can be discovered lurking within behavior.[16] A given instance of behavior can be interpreted as consistent with more than one purpose. For instance, suppose that a man awakens, hears suspicious sounds in his room, sees an image in the bedroom mirror, fires his gun, and shatters the mirror. What is the purpose manifested in this behavior? To shoot a burglar? To commit suicide? To break the mirror? Whereas Rosenblueth and Wiener emphasized behavior that many observers agree to be directed toward one specific goal, Taylor emphasized the diversity of purposeful accounts that can be adopted to explain an instance of behavior. Taylor conceded that the behavior itself provides the best evidence we have for judgments about purpose, but he maintained that this evidence falls far short of providing criteria for a dichotomous classification of behavior.

Moreover, Taylor continued, the value of the concept "purpose" is in question if the behaviors of rocking chairs, pendulums, compass needles, et al. qualify as "purposeful." The fact that such behaviors are "passive" rather than "active" does not save the concept from being too widely applicable to be useful. He asked, what is gained by thus calling tobacco pipes, rocking chairs, compass needles, weighted roulettes and so on, *purposeful*? What, for example, is science able to learn about these objects, which it might otherwise never discover, by regarding them in this light?[17]

Moreover, the Rosenblueth and Wiener characterization of purposeful behavior is not only too inclusive—it also is too exclusive. If we define purposeful behavior to be an achievement in which an object reaches "a definite correlation in time or in space with respect to another object or event,"[18] then certain types of *prima facie* purposeful behavior do not qualify as purposeful. Taylor cited cases in which the object of goal-directed behavior does not exist. On the Rosenblueth and Wiener account, neither the behavior of a man groping for the matches that earlier were on the table, nor the behavior of a boy searching a refrigerator for the piece of pie previously removed and eaten by his brother, would be instances of purposeful behavior. But surely these instances ought to count as purposeful sequences of actions. Taylor concluded that the vocabulary of science is not enriched by reintroducing the concept of purpose.

Ernest Nagel suggested in *The Structure of Science* (1961), that teleological explanations can be reformulated, without loss of content, as deductive-nomological explanations.[19] Consider the following teleological explanation: "Chlorophyll is present in green plants *in order that* photosynthesis take place." Nagel maintained that this statement is a telescoped deductive argument:

1. Green plants are supplied water, carbon dioxide, and sunlight, and they produce starch.
2. If green plants lack chlorophyll, even though they are supplied water, carbon dioxide, and sunlight, they do not produce starch.

∴ Green plants contain chlorophyll.

The teleological explanation indicates the *consequences* for the system of a constituent part. The allegedly equivalent deductive-nomological explanation states the *conditions* under which the system persists in its characteristic activities. The teleological account focuses on the *effect* of a given cause. Photosynthesis is the *effect* of the presence of chlorophyll (given that water, carbon dioxide, and sunlight are supplied). The deductive-nomological account focuses on a *cause* (necessary condition) of photosynthesis. But what difference does it make whether one focuses on the effect of a given cause or the cause of the same effect?

Photosynthesis does not qualify as "teleological purposeful behavior" on the Rosenblueth-Wiener-Bigelow taxonomy. The plant is not itself the source of the output energy in question. Nagel conceded this. However, he sought to extend analysis to typical cases of "active, goal-directed behavior." He called attention to the maintenance of temperature homeostasis in the human body. Despite wide fluctuations in the temperature of its environment, the body's internal temperature remains roughly constant. A number of physiological mechanisms are available to compensate for external temperature changes. Among these mechanisms are

1. dilation or contraction of peripheral blood vessels (which affects the rate of heat radiated or conducted by the skin);
2. thyroid gland activity (which affects the body's metabolic rate);
3. adrenal gland activity (which affects the rate of internal combustion);
4. muscular contractions (which generate internal heat); and
5. perspiration (in which heat is released by evaporation).

The organization of the body is such that if there is a "primary variation" (P) in one of these mechanisms, it is accompanied by an "adaptive variation" (A) in one or more of the other mechanisms. The result is that the condition of stasis (S) is maintained. Suppose that a "primary variation" in one of these mechanisms has occurred, and that the condition of stasis continues. The appropriate teleological explanation would be

A occurred, following P, in order that S.

Nagel maintained that this teleological explanation can be restated without loss of meaning as a deductive-nomological explanation:

$$(P \mathbin{\&} \sim A) \supset \sim S$$
$$P$$
$$S$$
$$\overline{}$$
$$\therefore A$$

Whereas the teleological account focuses on the *consequences* for the system of the adaptive variation, the deductive-nomological account focuses on the *conditions* requisite for maintenance of stasis.[20]

Nagel insisted that the deductive-nomological explanation does capture the distinctive character of this "goal-directed" system. It does so by specifying the conditions under which "directive organization" takes place. Nagel challenged those who believe that teleological interpretations have a "surplus meaning" to specify this meaning. If a teleological explanation does have a "surplus meaning," then it should be possible to cite evidence that supports the teleological account but not the corresponding deductive-nomological account. But no such evidence has been uncovered. Nagel insisted that so long as the available evidence indifferently supports both the teleological account and the deductive-nomological account, then we must conclude that the two accounts make the same assertions.[21]

Laws and Confirmation

The Nature of Lawlike Propositions

Writing in 1919, Norman R. Campbell declared that

> the object of science is the ordering of natural phenomena . . . A law is important for science because it represents the achievement of one stage in this ordering; it establishes a connection between a large number of previously disconnected observations.[1]

He held that a scientific law asserts that A is uniformly associated with B, where A and B are concepts that refer, in complex ways, to phenomena in the natural world.[2] Examples include:

1. The volume of a gas at constant pressure is directly proportional to its temperature (Charles).
2. The current in a closed circuit is proportional to the quotient of its voltage over its resistance.
3. The melting point of silver is 961.8°C.

Campbell noted that, in general, a law that stipulates a uniform association of two concepts can be established only by presupposing the truth of other laws. These other laws are involved in the instrumental operations used to measure values of the concepts.

Hooke's Law, for instance, states a proportionality between the force on a spring and its displacement ($F \propto \delta x$). To determine displacement, instrumental operations are required. Hooke's Law depends on the truth of the laws governing those operations. The simplest is to view the position of the end of the spring against a properly calibrated scale. A more precise measurement may be made by using an optical lever, the operation of which presupposes the truth of the law of refraction.

Laws that attribute properties to types of substances also presuppose the truth of other laws. For example, the law that "*A* is soluble in *B*" is established by appeal to instrumental procedures dependent on further laws. One may assess solubility (1) by taking weight-volume measurements from various points within *B*, or (2) by passing a beam of light through *B*.

The post-Second World War orthodox position was that scientific explanation is subsumption under general laws. But how is one to decide in a particular case whether a generalization qualifies as a law? Laws cannot be identified by logical form alone. Some generalizations are nomological and some are merely accidental.

We believe that "All sodium samples react with chlorine" is a law and that argument (1) below has explanatory force:

> 1. All sodium samples react with chlorine.
>   This sample is a sodium sample.
>   _____
>   ∴ This sample reacts with chlorine.

On the other hand, we believe that "All men now in this room are bald" is merely an accidental generalization, and that argument (2) below does not have explanatory force:

> 2. All men now in this room are bald.
>   Smith is a man now in this room.
>   _____
>   ∴ Smith is bald.

We adopt an attitude toward a nomological generalization that differs from our attitude toward an accidental generalization. We affirm that "if that sample were a sodium sample (although it is not), then it would react with chlorine." However, we do not affirm that "if that man now were in this room, then he would be bald." Orthodox theorists emphasized that nomological generalizations support contrary-to-fact conditionals, and accidental generalizations do not. But what is the nature of this support?

R. B. Braithwaite attributed this support to deductive relationships that link nomological generalizations to other laws. He declared that universal conditional *h* is lawlike if *h*

occurs in an established deductive system as a deduction from higher-level hypotheses which are supported by empirical evidence which is not direct evidence for *h* itself.[3]

The generalization about the chemical reactions of sodium is a deductive consequence of the postulates of the periodic table, which postulates are deductive consequences of the postulates of quantum theory. No such deductive relationships are known for the generalization about bald men now in this room.

Ernest Nagel extended Braithwaite's analysis. He called attention to several distinguishing characteristics of nomological generalizations.

In the first place, a nomological generalization has an open scope of predication. The generalization about the behavior of sodium samples, for instance, is applicable to samples over and above those that have been examined prior to a given time. Accidental generalizations, by contrast, often have closed scopes of predication. This is the case for the generalization about the men now in this room. An open scope of predication is one mark of nomological status.

Secondly, a nomological generalization is unlikely to restrict to a small region of space or time the individuals that satisfy the generalization.

Thirdly, as Braithwaite had emphasized, a nomological generalization often receives indirect support in virtue of its deductive relationships to other laws. For example, if generalizations L_1, L_2, and L_3 are deductive consequences of the postulates of a theory, then evidence that supports L_2 or L_3 also indirectly supports L_1. The generalization about sodium receives indirect support from evidence supporting similar generalizations about the reactions of lithium or potassium. Accidental correlations, on the other hand, do not receive this kind of indirect support.[4]

If Braithwaite and Nagel are correct, then the lawlike status of a generalization depends on its relations of deducibility to other statements. Consistent with this position, Nagel suggested that contrary-to-fact conditionals be interpreted as meta-linguistic claims about relations among other statements. To assert that "if *A* were the case (although *A* is not the case), then *B* would be the case" is to claim that *B* follows deductively from *A* in conjunction with suitable laws and statements about antecedent conditions. For example, consider the following contrary-to-fact conditional:

if the length of pendulum *P* were 1/4 its present length, then its period would be 1/2 its present period.

On Nagel's interpretation, the implicit meaning of this claim is that the following deductive relationship holds:

$$T \propto \sqrt{l}, \text{ for low amplitude swings of a simple pendulum}$$
under negligible air resistance.
$$A \text{ is a simple pendulum swinging at low amplitudes}$$
under negligible air resistance.
$$l_2 = 1/4\, l_1$$

$$\therefore T_2 = 1/2\, T_1 [5]$$

Orthodox theorists appealed to the hierarchical deductive structure of scientific theories in order to distinguish scientific laws from accidental correlations. They maintained that a universal conditional is lawlike if it is related deductively to other laws. Some of these other laws, in turn, may be supported indirectly by their logical relations to still further laws. But if the entire network of laws is to have empirical significance, at least some statements in science must be accepted as laws on the basis of *direct* evidential support.

The attempt to characterize lawlike universals thus leads to the problem of the confirmation of laws. Arthur Pap emphasized that a satisfactory analysis of the difference between lawlike universals and accidental universals can be given only in terms of a theory of confirmation. According to Pap, it is necessary to establish

> rules of critical inductive generalization in accordance with which the acceptance of a nomological implication is justified by "confirming" empirical evidence.[6]

Hempel's Program for the Evaluation of Hypotheses

A central claim of post-war orthodoxy is that a theory-independent observation language can be formulated. It was assumed that facts about the world can be expressed in this language, and that these facts are the ultimate court of appeal in the evaluation of hypotheses. An important task in the philosophy of science is to specify under what conditions, and to what extent, facts do support hypotheses.

In an essay published in 1945, Carl Hempel outlined a three-stage program for the evaluation of scientific hypotheses.[7] The first stage is to accumulate observation reports. The second stage is to establish whether an observation report confirms, disconfirms, or is neutral toward a given hypothesis. The third stage is to decide whether to accept a hypothesis given the available evidence.

Hempel focused primarily on the second stage of the evaluation process. He observed that hypotheses and observation reports are expressed as sentences within the language of science. An important problem in the philosophy of science is to uncover the logic of the relation between a hypothesis (*H*) and a statement recording evidence (*e*), if a suitable definition of "*e* confirms *H*" could be formulated, then one could apply it to decide whether a particular evidence statement confirms a given hypothesis.

The third stage in the evaluation of hypotheses involves judgments of a pragmatic nature. How much confirming evidence is needed before a hypothesis qualifies as acceptable? This depends, presumably, on judgments about the "explanatory power" and "simplicity" of the hypothesis, as well as judgments about the status of rival hypotheses. Hempel acknowledged the pragmatic nature of this phase of the evaluation process. Nevertheless, he suggested that it would be fruitful to reconstruct this phase in the terms of formal logic. This would require the selection of rules of acceptance and the development of a logic relating these rules to hypotheses and statements about confirming evidence.[8]

Many philosophers of science agreed that Hempel had correctly outlined an important research program for the philosophy of science. Hempel's essay contributed to the formulation of an "orthodox orientation" toward the problems of confirmation and acceptability. This orientation was widely shared, even by philosophers of science who did not accept Hempel's specific views on confirmation.

Hempel sought to work out the logic of the relation between hypotheses and observation reports. He presented his conclusions in two installments. The first installment is an analysis of alleged paradoxical consequences of the concept of confirmation. The second installment is the formulation of a definition of "*O* confirms *H*."

The Raven Paradox

In 1930, Jean Nicod had suggested the following relations between a scientific law and a statement that reports evidence:

Law	Evidence Statement	Relation
A entails B	A and B	Confirmation
	A and not-B	Invalidation

Nicod's criterion of confirmation stipulates that a law is supported by instances of a joint occurrence of *A* and *B*.[9] Tacit acceptance of this criterion has been widespread among scientists and philosophers of science.

Hempel pointed out that if a scientific law is expressed as a universal conditional — (x) (Ax ⊃ Bx)—then Nicod's table of relations may be expanded as follows:

Law	Evidence Statement	Relation
(x) (Ax ⊃ Bx)	Aa & Ba	Confirmation
	Aa & ~Ba	Disconfirmation
	~Aa & Ba	Neutral
	~Aa & ~Ba	Neutral

For the universal conditional "All ravens are black," (x) (Rx ⊃ Bx) is confirmed by an individual *a* just in the case *a* is a raven and *a* is black, and would be disconfirmed by an individual *b* which is a raven and is not black.

So far, so good. But it is a theorem of deductive logic that

$$(x) (Rx \supset Bx) \equiv (x) (\sim Bx \supset \sim Rx)$$

On Nicod's criterion, "All non-black things are non-ravens," (x) (~Bx ⊃ ~Rx) is confirmed by (~Ba & ~Ra) and is disconfirmed by (~Ba & Ra).

If we allow instances that confirm "All non-black things are non-ravens" also to confirm "All ravens are black" then the table of confirmation values is modified to:

Law	Evidence Statement	Relation
(x) (Ax ⊃ Bx)	1. Ra & Ba	Confirmation
	2. Ra & ~Ba	Disconfirmation
	3. ~Ra & Ba	Neutral
	4. ~Ra & ~Ba	Confirmation

It is perhaps counterintuitive that a non-black non-raven confirms the generalization that "All ravens are black." Moreover, the following relation also is a theorem of deductive logic:

$$(x) (Rx \supset B) \equiv (x) (\sim Rx \text{ v } Bx)$$

Since evidence statement "(~Ra v Ba)" is true provided that ~*Ra* is true, or *Ba* is true, or both, line (3) in the above table also counts as a confirming instance for "All ravens are black."

If we accept the equivalence condition "whatever confirms (disconfirms) one of two equivalent sentences also confirms (disconfirms) the other,"[10] then the table of confirmation values becomes:

Law	Evidence Statement	Relation
(x) (Ax ⊃ Bx)	1. Ra & Ba	Confirmation
	2. Ra & ~Ba	Disconfirmation
	3. ~Ra & Ba	Confirmation
	4. ~Ra & ~Ba	Confirmation

The table indicates that the law is confirmed not only by black ravens, but also by anything that is not a raven, black or otherwise. The possibilities for indoor ornithology are limitless. I examine every object in my study, ascertain that it is not a raven, and thereby confirm the law that all ravens are black.

Hempel emphasized that the "Raven Paradox" arises when four principles are affirmed. These principles are:

1. Nicod's criterion,
2. equivalence condition,
3. the assumption that many important scientific laws are universal conditionals properly symbolized by "(x) (Ax ⊃ Bx)", and
4. our intuitions about what should count as confirming evidence.

To dissolve the paradox, it is necessary to reject one or more of the four principles. Hempel advanced good reasons for retaining Nicod's criterion and the equivalence condition. He then examined two possible modifications of the third principle that interprets scientific laws to be universal conditionals.

The first modification is to tack on an existential rider. On this interpretation, "All ravens are black" is symbolized

(x) (Rx ⊃ Bx) & (∃ x) Rx.

Since this formula asserts the existence of ravens, instances of non-ravens do not confirm it.

The existential interpretation of scientific laws is effective in dissolving the Raven Paradox. However, the price paid is too great. Many important scientific laws are vacuously true. Consider the law of inertial motion. This law specifies the behavior of those bodies which are not under the influence of impressed forces. But no such bodies exist. And even if such a body did exist, we could have no knowledge of it. Observation of a body requires the presence of an observer or some recording apparatus. However, each body in the universe exerts a gravitational force on every other body. An observed body cannot be free of impressed forces.

The law of inertia specifies what happens in an idealized, non-realizable case. But by an appeal to this ideal case, scientists correlate progressively decreasing impressed forces with increasingly uniform rectilinear motion. Even though no force-free bodies exist, we do not want to formulate the law of inertia in such a way that it turns out to be false. And it would be false if we tack on an existential rider, such that

$$(x) (Fx \supset Rx) \ \& \ (\exists x) \ Fx$$

where $Fx = x$ is a moving body not subject to an impressed force, and
          $Rx = x$ is a body that continues in undiminished rectilinear motion.

Other vacuously true laws are Galileo's law of falling bodies and Archimedes' law of the lever. In general, those laws that specify what happens in idealized circumstances—e.g., free-fall in a vacuum, the balancing of weights hung from a mass-less but infinitely rigid rod—are vacuously true. To formulate them as making existential claims is to render them false.

It does violence to important scientific practice to interpret scientific laws as making existential claims. This is one good reason to reject this modification of principle (3).

A second reason to reject adding an existential rider to the expression of a scientific law is that scientists often take a universal conditional to be logically equivalent to its contrapositive. Hempel called attention to the use of flame tests in inorganic chemistry.[11] Scientists take the following universal conditionals to be equivalent:

(1) $(x) (Bx \supset Gx)$

(2) $(x) (\sim Gx \supset \sim Bx)$

where $Bx = x$ is a case in which a barium-containing salt is exposed to a
          flame, and
          $Gx = x$ *is* a case in which a green flame is present.

A scientist might synthesize a new barium compound and perform a flame test to confirm relation (1). On the other hand, a scientist might perform a flame test on a compound of unknown composition, observe that the flame color is not green, and use relation (2) to conclude that no barium is present in the compound. He might then perform chemical analyses on the compound to establish by that means that no barium is present. Since the strategy of tacking on an existential rider would destroy the equivalence of (1) and (2), this strategy is contrary to scientific usage of the above type.

Moreover, if scientific laws are reformulated to include existential claims, in some cases it is difficult to know that to which the existential clause is to refer. An example given by Hempel is

> in all cases, if a person receives injection *I* and has a positive skin reaction, then that person has diphtheria.

It is not clear from the wording of the law what claim about existence should be made. Is it that "persons exist"? Or that "there exist persons who have received injection *I*"? Or that "there exist persons who have displayed a positive skin reaction after receiving injection *I*"? Hempel concluded that the disadvantages of an existential interpretation of laws are too great to justify taking this escape from the Raven Paradox.[12]

A second modification of principle (3) discussed by Hempel is to restrict the range of application of a scientific law to the class mentioned in the antecedent clause. On this view, "All ravens are black" is interpreted to mean "(x) $(Rx \supset Bx)$ (class: ravens)."

This modification does not assign existential meaning. Hence it does not render false vacuously true laws. Moreover, by restricting confirming instances to the class of ravens, the Raven Paradox is avoided.

However, this modification too is open to objections. In theoretical physics, a given formula is subject to replacement by logically equivalent forms without regard to any change in "range of application." The "range of application" modification is contrary to important scientific practice. It also is subject to the previously mentioned objections based on the flame test and diphtheria examples.

Hempel maintained that uses of Nicod's criterion and the equivalence condition are deeply embedded in scientific practice, and that many scientific generalizations are translated correctly as universal conditionals. His own position on the Raven Paradox is that we are misled by our intuitions.

He pointed out that the generalization "All ravens are black" is not exclusively about ravens. Since "(x) $(Rx \supset Bx)$" is logically equivalent to "(x) $(\sim Rx \vee Bx)$," the raven generalization is really about all the entities in the universe. It states that "for all entities in the universe, either that entity is not a raven or it is black."

In addition, our intuitions about confirmation mislead us because we make implicit judgments about relative class sizes. We know that the class of ravens is very much smaller than the class of non-black entities. The likelihood of finding a disconfirming instance of the raven generalization is much greater if we survey ravens for color than if we examine non-black items for ravenhood.

Suppose, however, that our background knowledge was different. If we knew that the universe contained a million times more ravens than non-black entities, then it would be prudent to examine non-black entities for ravenhood rather than to examine ravens for color.

Consider once again the flame test for barium. Given that

$$(x) (Bx \supset Rx),$$

we sometimes hold that the instance (~Ba & ~Ra) confirms the generalization. We count as a confirming instance an unknown sample that does not give a green flame test, provided that the absence of barium subsequently is established on chemical grounds. But we do not count as a confirmation instance a negative test on an ice cube. This is because we knew in advance that no barium was present. The sample was not unknown with regard to the presence of barium. But if we had no antecedent knowledge of the chemical composition of ice, then there would be no difference in the two cases.

Hempel's position is that the relationship between generalizations and their confirming instances is not paradoxical to the properly educated intuition:[13] that "All ravens are black" is confirmed by finding black ravens, non-black non-ravens, and black non-ravens. If one keeps in mind the logical form of a universal generalization, and if one excludes background knowledge about relative class sizes, qualitative confirmation is not a paradoxical relation.

The "Satisfaction Theory" of Qualitative Confirmation

Hempel sought to formulate a definition of qualitative confirmation that would be effective in the context of the evaluation of scientific hypotheses. His strategy was to create an artificial language and specify a set of conditions necessary and sufficient for the definition of "*O* confirms *H*" in that language. The content of that language includes:

terms denoting individuals—a, b, c, . . .;
property terms—Ax, Bx, Cx, . . .;
relational terms—Rxy, Sxyz, . . .;
logical connectives—&, v, ⊃, ≡, . . .;
quantifiers—(x) . . ., (∃ x) . . .;
rules of sentence formation;
rules of deductive inference;
observation sentences—Aa, Rbc, . . .;
hypotheses—any properly formed sentence within the language.

Hempel listed four requirements that a definition of "O confirms H" must satisfy:

1. Entailment condition—If O entails H, then O confirms H.
2. Equivalence condition—If O confirms H, then O confirms every H^* logically equivalent to H.
3. Special consequence condition—If O confirms H, then O confirms every logical consequence of H.
4. Consistency condition—O is logically compatible with the class of all hypotheses that it confirms.[14]

Hempel formulated a definition of "O confirms H" that satisfies the above four conditions. The formulation proceeds in stages from the concept of the "development of H" to the relation "O directly confirms H" to the relation "O confirms H."

The "development of H" for a finite set of objects O_n asserts what H would assert if no objects other than these objects existed.[15] The "development" of the raven hypothesis for the set of objects $[a, b, c]$ is "[(Ra ⊃ Ba) & (Rb ⊃ Bb) & (Rc ⊃ Bc)]." Hempel stipulated that observation report O "directly confirms" hypothesis H if O entails the development of H for the set of objects mentioned in O. [16]

Suppose O = [(Ra & Ba) & (~Rb & Bb) & (~Rc & ~Bc)]. O "directly confirms" H because

$$Ba \rightarrow (Ba \vee \sim Ra) \rightarrow (Ra \supset Ba)$$
$$Bb \rightarrow (Bb \vee \sim Rb) \rightarrow (Rb \supset Bb)$$
$$\sim Rc \rightarrow (\sim Rc \vee Bc) \rightarrow (Rc \supset Bc)$$

This conclusion is consistent with Hempel's position on the Raven Paradox. "All ravens are black" is "directly confirmed" by instances of black ravens, instances of black non-ravens, and instances of non-black non-ravens.

Hempel stipulated further that

O* confirms H if H is entailed by a class of sentences each of which is directly confirmed by O*.[17]

Suppose O* = (~Rd & Bd). O* confirms H = (x) (Rx ⊃ Bx) because O* is a member of the class [(Ra ⊃ Ba), (Rb ⊃ Bb), (Rc ⊃ Bc), (Rd ⊃ Bd), each member of which entails the development of H for the class $[a, b, c, d]$., viz., [(Ra ⊃ Ba) & (Rb ⊃ Bb) & (Rc ⊃ Bc) & (Rd ⊃ Bd)]. Hempel declared that

the criterion expressed in these definitions might be called the *satisfaction criterion of confirmation* because its basic idea consists in construing a hypothesis as confirmed by a given observation report if the hypothesis is satisfied in the finite class of those individuals which are mentioned in the report.[18]

Rudolf Carnap argued that the conditions that Hempel imposed on an explication of "*O* confirms *H*" were inadequate. He produced a counter-case to the special consequence condition, and noted that the entailment condition was not adequate to cover cases in infinite language systems.[19] Carnap noted, moreover, that Hempel held that the consistency condition implied both:

1. If *O* and *H* are logically inconsistent, then *O* does not confirm *H*.
2. If *H* and *H** are logically inconsistent then *O* confirms neither.

Requirement (2) is violated by cases in which *H* and *H** assign slightly different relative frequencies to the incidence of a property within a population. Suppose, for instance, that *H* states that 87 percent of the individuals in a population possess trait *T*, whereas *H** states that the incidence of *T* in the population is 88 percent. An observation report that reveals that 875 individuals in a sample of 1000 from that population have *T* would confirm both *H* and *H**. And yet *H* and *H** are logically inconsistent.[20] Since the consistency condition implies (2), it cannot be retained. Hempel conceded in a "Postscript (1964) on Confirmation" that Carnap's criticisms were well-founded.[21] The satisfaction criterion of qualitative confirmation is not satisfactory.

Carnap on Quantitative Confirmation

There was general agreement among philosophers of science that it would be desirable to develop a quantitative measure of confirmation. It is of some value to know that *e* confirms *H*. It would be far more useful to know that *e* confirms *H* to degree *c*.

In *The Logical Foundations of Probability* (1950), Rudolf Carnap outlined a theory of "degree of confirmation."[22] His approach was to

1. formulate an artificial language,
2. define "c (e, H)" for the sentences of that language, and
3. show that values calculated from the definition in specific cases agree with our intuitive impressions about the degree of evidential support *c* that *H* receives from *e*.

The ingredients of Carnap's artificial language include:

1. truth-functional connectives and quantifiers,
2. individual constants that name individuals,
3. primitive predicates, and
4. rules of sentence formation and deductive inference.

Carnap placed three requirements on the predicates of the language:

1. They are finite in number.
2. They are coordinate. If P and Q are coordinate predicates, and if P and Q represent equally possible alternatives, then if $Q = (R \vee S)$, neither $(P \vee R)$ nor $(P \vee S)$ represent equally possible alternatives. For instance, take $P =$ purple, $Q =$ non-purple, $R =$ red, and $S =$ sepia. If the members of the pair [purple, non-purple] represent equally possible coordinate properties, then neither the members of the pair [purple, red] nor members of the pair [purple, sepia] represent equally possible properties.
3. They are logically independent of one another. The requirement of logical independence restricts the language to one predicate for each determinable, In the case of the determinable "color," for example, if "red" is included, "green" is excluded. since "*a* is red" implies "*a* is not green." The requirement of logical independence also excludes quantitative predicates, which take on a series of values relative to some scale. Carnap was aware that his artificial language was impoverished in these respects, and in subsequent studies extended his investigations to languages that include families of related properties.[23]

Given a finite set of predicates, and a finite set of individual constants that designate univocally the individuals of the system, the system is fully described by a particular state-description. A state-description attributes each property, or its negation, to each of the individuals in the system. For a simple system with two individual constants a and b, and two property terms P and Q, there are 16 possible state-descriptions, one of which must be actualized, viz.:

State-Descriptions for a Two-Individual-Two-Property System

1. Pa & Qa & Pb & Qb	9. ~Pa & Qa & Pb & Qb
2. Pa & Qa & Pb & ~Qb	10. ~Pa & Qa & Pb & ~Qb
3. Pa & Qa & ~Pb & Qb	11. ~Pa & Qa & ~Pb & Qb
4. Pa & Qa & ~Pb & ~Qb	12. ~Pa & Qa & ~Pb & ~Qb
5. Pa & ~Qa & Pb & Qb	13. ~Pa & ~Qa & Pb & Qb
6. Pa & ~Qa & Pb & ~Qb	14. ~Pa & ~Qa & Pb & ~Qb
7. Pa & ~Qa & ~Pb & Qb	15. ~Pa & ~Qa & ~Pb & Qb
8. Pa & ~Qa & ~Pb & ~Qb	16. ~Pa & ~Qa & ~Pb & ~Qb

Carnap stipulated that each state-description in a language containing a finite number of individual constants receives from "tautological evidence" a degree of confirmation greater than zero.

Since a tautology such as "Pa v ~Pa" can provide no evidence that anything is the case, Carnap's stipulation is tantamount to an assignment of an initial degree of confirmation to each state-description, i.e., $c_0 (S) > 0$.

Carnap resolved to interpret "c (H, e)" as a measure of the logical probability of *H*, given *e*. He proved that if one adopts the stipulation above, and applies the axioms of the calculus of probabilities to the function "c (H, e)," then it follows that

$$c (H, e) = c_0 (H, e) / c_0 (e),$$

where $c_0 (H, e)$ is the initial confirmation of *e*.[24] He also proved that, for languages with infinitely many individual constants, if c (H, e) converges as the number of individuals *n* increases, then

$$c (H, e) = \lim c_0 (H, e) / c_0 (e).^{[25]}$$

$$n \rightarrow \infty$$

These are provocative results. They show that values of "degree of confirmation," interpreted as a logical probability, depend exclusively on how initial values are assigned. The next task, then, is to select an initial assignment of values to the statements of the language so that calculated values of c (H, e) agree with our intuitive expectations.

One possibility for assigning initial degrees of confirmation would be to give equal initial degrees of confirmation to each state-description within the language. For the "two-individual-two-property system" above, each state-description would be assigned $c_0 (S) = 1 / 16$, so that the initial degree of confirmation for the conjunction of all state-descriptions is 1. For hypothesis H = Pb & Qb and evidence statement e = Pa & Qa,

$$c (H, e) = c_0 (H, e) / c_0 (e) = 1 / 16 / 4 / 16 = 1 / 4$$

$c_0 (H, e) = 1 / 16$ because "Pb & Qb & Pa & Qa" holds in just one state-description (line 1), and $c_0 (e) = 4 / 16$ because "Pa & Qa" holds in four state-descriptions (lines 1, 2, 3, and 4)

However, $c_0 (H, e) = 4 / 16 = 1 / 4$ as well. Thus the degree of confirmation of "Pb & Qb," given that "Pa & Qa," is the same as the initial degree of confirmation of "Pb & Qb" itself. If equal initial degrees of confirmation are assigned to state-descriptions, then the accumulation of evidence is irrelevant to the determination of confirmation values. This result does not accord with our intuitions about confirmation.[26]

For this reason, Carnap chose to assign equal initial degrees of confirmation, not to state-descriptions, but to "structure-descriptions." A structure-description is a disjunction of isomorphic state-descriptions. Two state-descriptions are isomorphic if they differ only in the individuals for which a given set of properties is assigned. Given the schema [P & ~Q & ~P & Q], state-descriptions (1) and (2) are isomorphic:

1. Pa & ~Qa & ~Pb & Qb
2. Pb & ~Qb & ~Pa & Qa

The disjunction (1) v (2) is a structure-description within the two-individual-two-property system. There are 10 structure-descriptions within this language, viz.:

Structure-Descriptions	Number of State-Descriptions Disjoined
1. Pa& Qa & Pb & Qb	1
2. [Pa& Qa & Pb & ~Qb] v [Pb & Qb & Pa & ~Qa]	2
3. [Pa& Qa & ~Pb & ~Qb] v [Pb & Qb ~Pa & Qa]	2
4. [Pa& ~Qa & Pb & Qb] v [Pb & ~Qb & Pa & Qa]	2
5. [Pa& ~Qa & Pb & Qb] v [Pb & ~Qb & Pa & Qa]	2
6. Pa& ~Qa & & Pb & ~Qb	1
7. [Pa& ~Qa & ~Pb & Qb] v [Pb & ~Qb & ~Pa & Qa]	2
8. [Pa& ~Qa & ~Pb & ~Qb] v [Pb & ~Qb & ~Pa & ~Qa]	2
9. ~Pa& Qa & ~Pb & Qb	1
10. ~Pa& ~Qa & ~Pb & ~Qb	1

Carnap's formula for the initial degree of confirmation of state-description S is $c_0 (S) = 1 / n$ i, where n is the number of structure-descriptions in the language, and i is the number of state-descriptions isomorphic with S. For the language under consideration, Carnap would assign an initial degree of confirmation of $1 / 10$ to each structure-description above. For each state-description disjoined on lines 2, 3, 4, 5, 7, and 8, $c_0 (S) = 1 / 10 \times 2 = 1 / 20$. For each state-description on lines 1, 6, 9, and 10, $c_0 (S) = 1 / 10 \times 1 = -1 / 10$. Thus not every state-description has the same initial degree of confirmation.

The assignment of equal initial degrees of confirmation to structure-descriptions emphasizes uniformity. The degree of confirmation of a hypothesis that an individual has a particular combination of properties is increased by evidence that other individuals possess this same combination of properties.

Consider once again hypothesis H = Pb & Qb and evidence statement e = Pa & Qa.

Sentence	State-Descriptions in which the sentence holds	$c_0 (S) = 1 / n_i$	c_0
H = Pb & Qb	1	1 / 10	5 / 20
	5	1 / 20	
	9	1 / 20	
	13	1 / 20	
e = Pa & Qa	1	1 / 10	5 / 20
	2	1 / 20	
	3	1 / 20	
	4	1 / 20	
(H & e) = [Pb & Qb & Pa & Qa]	1	1 / 10	1 / 10

For the above example,

$$c [(Pb \& Qb), (Pa \& Qa)] = c_0 [(Pb \& Qb), (Pa \& Qa)] / c_0 (Pa \& Qa)$$
$$= 1 / 10 / 5 / 20 = 2 / 5.$$

Since c (H, e) = 2 / 5 is greater than c_0 (H) = 1 / 4, the hypothesis is strengthened by the accumulation of evidence that a second individual has the properties in question.

Thus far, the assignment of equal initial degrees of confirmation to structure-descriptions seems to accord with our intuitions about confirmation. However, there are other consequences of Carnap's assumptions that do seem to violate our intuitions. In particular, there are difficulties with universal generalizations of the form (x) (Px ⊃ Qx). For a language in which infinitely many substitution instances are possible, the degree of confirmation of a universal generalization is zero. And for languages with a large, but finite, number of substitution instances, the degree of confirmation of a universal generalization is very close to zero.

Carnap acknowledged that these theorems derived in his artificial language are *prima facie* counterintuitive.[27] He conceded that scientists regard as "well-confirmed" universal generalizations for which there is extensive positive evidence and no negative evidence. He pointed out, however, that when a scientist uses a universal generalization in an engineering application, or for purposes of prediction, he need not commit himself to the truth of the generalization over a large number of instances. It suffices that the generalization hold true in the next instance (or several instances). Consider an engineer's use of a physical law in the design of a bridge, Carnap maintained that

> when he says that the law is very reliable, he does not mean to say that he is willing to bet that among the billion of billions, or an infinite number, of instances to which the law applies there is not one counterinstance, but merely that this bridge will not be a counterinstance, or that among the bridges that he will construct in his lifetime, there will be no counterinstance. Thus *h* is not the law *l* itself but only a prediction concerning one instance or a relatively small number of instances. Therefore, what is vaguely called the reliability of a law is measured not by the degree of confirmation of the law itself but by that of one or several instances.[28]

Thus, according to Carnap, it is not cause for concern that c (H, e) = 0 for universal generalizations in infinite universes of discourse. The "instance confirmation" of a law is more important. The instance confirmation of a law—c_i (L, e)—is the degree of confirmation, on evidence *e*, that a new instance not mentioned in *e* fulfills the law. Indeed, what counts, for practical purposes, is the "qualified instance confirmation" of the generalization. The qualified instance confirmation of a generalization of the form (x) (Px ⊃ Qx) is the degree of confirmation, on evidence *e*, that a new instance not mentioned in *e* but which satisfies antecedent condition *P* also satisfies consequent condition *Q*. Carnap proved that c_{qi} (P, Q, e) approaches the value 1.0 as sample size increases, provided only that the number of observed negative instances is zero (or a fixed small number).[29]

Hence, although Carnap's theory suggests a degree of confirmation of zero for "All ravens are black," it also yields a qualified instance confirmation of 1.0. And since we operate in scientific contexts with a rudimentary concept of "qualified instance confirmation," the theory is not really counterintuitive after all. Carnap remarked at the end of *Logical Foundations of Probability* that

> the system of inductive logic here proposed ... is intended as a reconstruction restricted to a simple language form of inductive thinking as customarily applied in every day life and in science.[30]

Theories as Interpreted Axiom Systems

Campbell on Hypotheses, Dictionaries, and Analogies

Theories are on the highest rung of the ladder of language levels in science. Orthodox theorists were particularly interested in the relationship of theories to the lower rungs of the ladder. They believed that this relationship was the key to the "problem of theoretical terms," one aspect of which is to specify how claims about electrons and genes can be cashed on the observational level.

Post-war analyses of the relationship were based on a distinction between an axiom system and its application to experience. This distinction had been implied in Newton's formulation of mechanics. The importance of the distinction had been highlighted by the invention of noneuclidean geometries in the nineteenth century. And the role of the distinction in theory-construction had been analyzed by Norman Robert Campbell in *Foundations of Science* (1919).[1] Campbell's work is the point of departure for post-war discussions of scientific theories.

Campbell proposed a "hypothesis-plus-dictionary" view of the structure of scientific theories. It is based on a distinction between an abstract realm and an empirical realm. The hypothesis is a collection of statements in the abstract realm that are "incapable of proof or disproof by themselves."[2] The dictionary provides an empirical interpretation of this abstract realm. Dictionary entries link terms that occur in the statements of the hypothesis to observed properties in the empirical realm. The hypothesis-plus-dictionary view may be represented as shown in Figure 6.

α, β, υ, and δ are terms of the hypothesis. The lines joining these terms represent the basic statements (axioms) of the hypothesis. O_1 and O_2 are observables whose values are subject to measurement.

As a theory of procedure, the hypothesis-plus-dictionary view places emphasis on the creative imagination of the scientist. She may construct a simple, or a complex, abstract structure, subject only to the requirement that dictionary

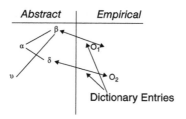

Figure 6 The Hypothesis-Plus-Dictionary View of Theories.

entries be provided that link the structure to empirically determined magnitudes. Whether or not the resulting theory is successful depends on its agreement with observations.

This is not a new interpretation of the structure of theories. It had been stated clearly by Pierre Duhem in *The Aim and Structure of Physical Theory* in 1906.[3] Some 220 years earlier Isaac Newton had introduced a distinction between abstract propositions and their empirical interpretation. Newton distinguished the "mathematical principles of natural philosophy" from the associated empirical laws that are subject to experimental confirmation.

The "mathematical principles" include the three axioms of motion—the principle of inertia, the proportionality of force and acceleration, and the equality of action and reaction. Newton maintained that these principles specify the motion of bodies in absolute space and absolute time. Neither distance in absolute space nor intervals of absolute time can be measured directly. What is needed are dictionary entries that link absolute values of space and time with their "sensible measures." To that end Newton specified procedures to convert motions in absolute space and time to motions within an empirically measurable coordinate system.

Newton provided a straightforward dictionary entry for absolute space. He recommended that the center of mass of the solar system be taken as the center of absolute space. Coordinate systems for the empirical determination of distances then may be constructed upon this central point.

Newton's suggested dictionary entry for absolute time is more complex. He did not specify any one periodic process as a measure of absolute time. He suggested instead that various processes for measuring temporal intervals be compared. Time may be measured by the flow of water through a hole in the bottom of a bucket, by the swings of a pendulum, by the motion of the earth around the sun, etc. Given a set of empirical procedures for measuring time, the determination of time-dependent sequences such as bodies in free-fall will be more regular under some procedures for measuring time than under others. Newton suggested that the best sensible measure of time is the procedure that yields the greatest regularity in the confirmation of empirical laws. On the hypothesis-plus-dictionary view, Newton's theory of mechanics may be represented as shown in Figure 7.

"Mathematical Theories" and "Mechanical Theories"

Campbell drew a distinction between "mathematical theories" and "mechanical theories." Both types of theory have the hypothesis-plus-dictionary structure. In a mathematical theory, each term mentioned in the abstract realm is correlated

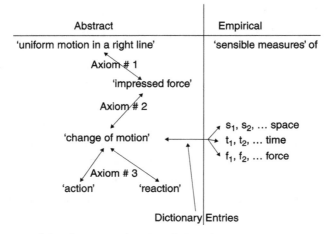

Figure 7 Newton's Mechanics on the Hypothesis-Plus-Dictionary View.

Source: John Losee, *A Historical Introduction to the Philosophy of Science*, 4th edn. (Oxford: Oxford University Press, 2001), 80.

with a dictionary entry that links it to an observable whose values may be measured. In a mechanical theory, some, but not all, terms of the hypothesis are linked to the empirical realm by a dictionary entry. Campbell noted that

> there is obviously a great difference between a theory in which some proposition based on experiment can be asserted about each of the hypothetical ideas, and one in which nothing can be said about these ideas separately, but only about combinations of them.[4]

Mathematical Theories

Physical geometry is a mathematical theory. The hypothesis-plus-dictionary view provides a useful interpretation of the distinction between pure geometry and physical geometry. Pure geometry remains in the abstract realm. Physical geometry makes claims that can be tested empirically. Consider, for instance, the theory about triangles shown in Figure 8.

Hypothesis

undefined terms—"point," "line"
defined terms—"triangle," "apex," midpoint," "intersection"
axioms—"A straight line may be drawn between any two points," et al.

theorem— "The line drawn from each of the three apices of a triangle to the midpoint of the opposite side intersect at a point within the triangle," viz.:

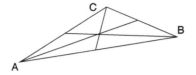

Figure 8 A Theory about Triangles.

Dictionary Entries

"point" corresponds to a dot on paper created by pressing a sharp pencil upon the paper

"line" corresponds to the result of an operation drawing a pencil along a straightedge placed on paper

"midpoint" is established by operations with compass and straightedge

"intersection" is a point created by crossing lines

"triangle" is a plane figure enclosed by three lines which meet at three points

Each term of the hypothesis—"point," "line," "triangle," "apex," "midpoint," and "intersection"—is given an empirical correlate. The resultant theory (hypothesis-plus-dictionary) is an empirical claim of physical geometry. It is subject to empirical confirmation, or refutation, *qua* scientific law. Physical geometry as a whole is a Campbellian mathematical theory. This is the case regardless of the axiom system of pure geometry selected—e.g., those of Euclid, Riemann, or Lobatchevsky.

Campbell cited Fourier's theory of heat conduction as an example of a mathematical theory. This theory displays the hypothesis-plus-dictionary structure. The hypothesis is a differential equation applied to an infinitely long slab of material whose density, specific heat, and thermal conductivity remain to be specified, viz.

$$\lambda \,(\partial^2 \theta / \partial x^2 + \partial^2 \theta / \partial y^2 + \partial^2 \theta / \partial z^2) = \rho \, c \,(\partial \theta / \partial t)$$

where λ is thermal conductivity, ρ is density, c is specific heat, θ is absolute temperature, t is time, and $x, y,$ and z are Cartesian coordinates of a point within an infinite slab of material.

Dictionary entries are provided for each of the terms in the equation for the material under investigation. The result is an empirical law that specifies the rate of heat conduction within the material in question. The empirical law has the same mathematical form as the abstract hypothesis.

Mechanical Theories

Mechanical theories provide dictionary entries for some, but not all, of the terms in the hypothesis. Newton's mechanics, Mendel's theory of heredity, the kinetic theory of gases, and Bohr's theory of the hydrogen atom all are mechanical theories.

Mendel's theory fits the hypothesis-plus-dictionary format (see Figure 9). Gregor Mendel (1822–1884) studied the transmission of pod color in pea plants (*genus pisum*). Pea plants produce either green pods or yellow pods. They do not produce pods of an intermediate yellow-green color.

Mendel crossed a green-pod plant with a yellow-pod plant. The first generation contained both green-pod plants and yellow-pod plants. He then crossed two green-pod plants from the first generation, creating a hybrid cross.

In Mendel's theory there is no dictionary entry for the concept "recessive gene." In the case of pea plants the presence of the recessive gene in a hybrid cross may be associated with either green-pod color or yellow-pod color. Nevertheless, the hypothesis as a whole is linked to observable pod colors.

The kinetic theory of gases also may be recast in the hypothesis-plus-dictionary format (see Figure 10).

The axioms of the theory state claims about the behavior of molecules that comprise the gas. The motions of these molecules are correlated—*via* dictionary entries—with the pressure and temperature of the gas. Among the theorems implied by the theory are the laws attributed to Boyle $(P \supset 1 / V)_{m, T = k}$, Charles $(V \supset T)_{m, P = k}$, and Gay-Lussac $(P \supset T)_{m, V = k}$.

The kinetic theory of gases includes dictionary entries that link u, the root-mean-square velocity of all the molecules, to temperature and pressure. No dictionary entry is provided for the velocity of an individual molecule, a basic concept of the hypothesis.

Niels Bohr's theory of the hydrogen atom (1913) attracted much attention at the time it was formulated. It postulated a set of circular orbits for the hydrogen electron with transitions between orbits accompanied by discrete, or "quantized," changes of energy (see Figure 11).

Values of n and λ calculated in the abstract realm are correlated—*via* dictionary entries—with lines observed in the hydrogen spectrum. The intensity and spacing of major lines in the observed spectrum match predictions drawn from the theory.

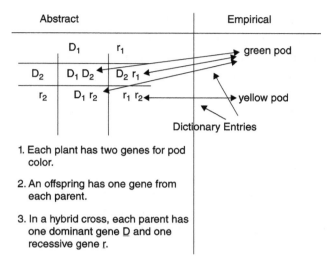

Figure 9 Mendel's Theory on the Hypothesis-Plus-Dictionary View.

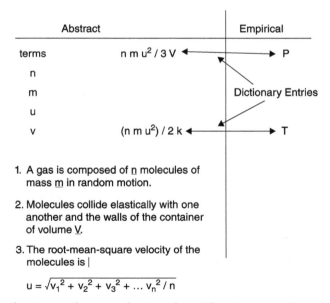

Figure 10 The Kinetic Theory on the Hypothesis-Plus-Dictionary View.

Bohr's theory provides no dictionary entry for the location of the electron, whether it is in a circular orbit or between orbits in an energy transition. More troubling is the ambivalent treatment of classical electromagnetic theory. Bohr applied Coulomb's Law to the motion of an electron in its orbit. However, he denied that the circulating electron radiates energy. In classical electromagnetic theory a charged particle revolving around a central point does radiate energy.

The Role of Analogy

According to Campbell, the hypothesis-plus-dictionary view of theories presented above is incomplete. To show this, Campbell formulated a theory to account for the temperature dependence of electrical resistance in pure metals (see Figure 12).

Given axiom "x = y," one may deduce that "a x^2 + a y^2 = 2a y^2" and that "a (x^2 + y^2) = 2a y^2 b / b," and finally that

$$a (x^2 + y^2) = 2 a b (x y / b).$$

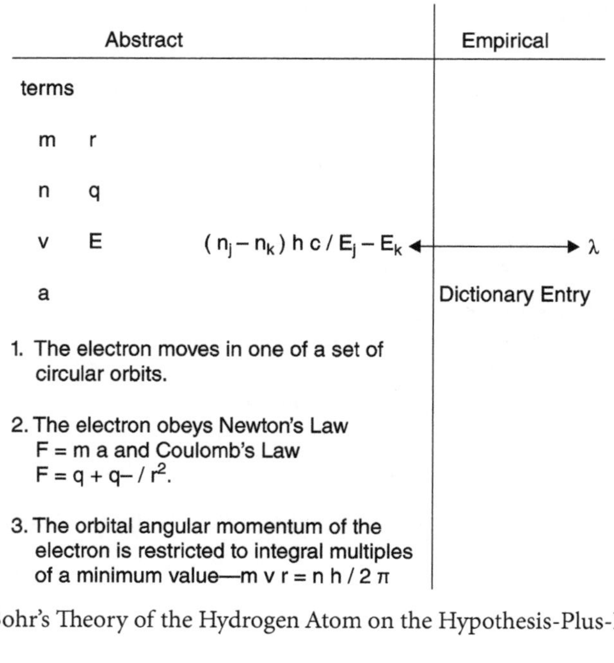

Abstract			Empirical
terms			
m	r		
n	q		
v	E	$(n_j - n_k)$ h c / $E_j - E_k$ ←	→ λ
a			Dictionary Entry

1. The electron moves in one of a set of circular orbits.

2. The electron obeys Newton's Law F = m a and Coulomb's Law F = q + q– / r^2.

3. The orbital angular momentum of the electron is restricted to integral multiples of a minimum value—m v r = n h / 2 π

Figure 11 Bohr's Theory of the Hydrogen Atom on the Hypothesis-Plus-Dictionary View.

According to the dictionary, this last expression states that the electrical resistance of a piece of pure metal is directly proportional to its absolute temperature. Campbell noted that "any fool can invent a logically satisfactory theory to explain any law".[5]

Campbell's artificial theory resembles the kinetic theory of gases in an essential respect. The hypothesis-plus-dictionary of each theory implies a well-confirmed empirical law. However, Campbell, unlike Maxwell, did not receive credit for the creation of an important scientific theory.

According to Campbell, the reason for this is that the kinetic theory features an analogy and the "theory" of the temperature dependence of electrical resistance in metals does not. Campbell required that every successful scientific theory display an analogy. He declared that

> in order that a theory may be valuable it must . . . display an analogy . . . Analogies are not "aides" to the establishment of theories; they are an utterly essential part of theories, without which theories would be completely valueless.[6]

Gerd Buchdahl noted that Campbell stressed the essential role of analogy in order to

> emphasize (indirectly) the intrinsic untestability of hypotheses . . . to insist that the *meaning* which such hypotheses have for us (if any) is due to the analogy *qua* analogy alone.[7]

Mechanical Theories and Analogies

Each of the mechanical theories discussed above displays an analogy. Newton's mechanics involves appeal to analogies. There are intimations of force, resistance, reaction, and inertial motion in everyday experience. We are conscious of

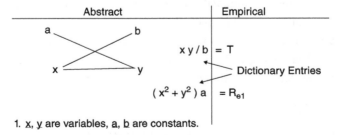

1. x, y are variables, a, b are constants.

2. x = y.

Figure 12 A Theory about the Temperature Dependence of the Electrical Resistance of Pure Metals.

exerting force in our activities and observing its effects. We have been involved in collisions and have experienced reactions to them. Moreover, we have experienced a relation between motion and resistance. In crude terms, the greater the resistance, the slower the motion, for a given degree of effort. The notion of inertial motion is plausible as an extrapolation to the case of zero resistance.

Mendel's theory—as does Darwin's—cites an analogy to processes in which material particles persist unchanged. Genes persist unchanged in reproduction, just as atoms persist unchanged in chemical reactions. For instance, a recessive gene in a green-pod pea plant retains its identity in a hybrid cross just as a sodium atom retains its identity (*qua* ion) in a reaction that produces salt.

The kinetic theory of gases invokes an analogy to collisions between material bodies. We are encouraged to superimpose on the hypothesis-plus-dictionary the picture of three-dimensional collisions of miniature billiard balls. Campbell declared that

> the theory of gases explains Boyle's Law, not only because it shows that it can be regarded as a consequence of the same general principles as Gay-Lussac's Law, but also because it associates both laws with more familiar ideas of motions of elastic particles.[8]

Bohr's 1913 theory of the hydrogen atom often is labeled the "planetary model" of the atom. The analogy associated with the theory is that just as planets move in stable orbits around the sun, so also does the electron move in one of a set of stable orbits around its proton.

The Copenhagen interpretation of quantum theory, developed by Niels Bohr and Werner Heisenberg (1926–1927), postdates Campbell's presentation of the hypothesis-dictionary-analogy view of theories. The Copenhagen interpretation fulfills the requirements Campbell placed on "mechanical theories." However, its analogy is unusual in certain respects.

The hypothesis of quantum theory may be expressed either in the matrix mechanics of Heisenberg or in the equation of Schrodinger. The Schrodinger equation for the state of a quantum system in spatial dimension x at time t is

$$\Psi_x = A\ e^{2\ \pi\ \mathrm{I}\,\mathrm{h}\,(\mathrm{Px}\,=\,\mathrm{Et})}$$

where A is a constant, e is the base of the system of natural logarithms, i is $\sqrt{-1}$, h is Planck's constant, P is a momentum operator, and E is an energy operator.

Since the Schrodinger equation contains an imaginary number—$\sqrt{-1}$—the state function Ψ has no empirical correlate. Max Born suggested that, since Ψ^2 is a real number, it may be linked to observables. Born introduced dictionary entries that link $\Psi\Psi^*$ dV to electron charge densities, orbital transition frequencies, and scattering distributions. In the expression Ψ Ψ^* dV, Ψ^* is the complex conjugate formed by substituting $\sim i$ for i in function Ψ. This expression usually is represented as $|\Psi^2|$ dV, with the vertical lines indicating the product of Ψ and its complex conjugate Ψ^*.

The Copenhagen interpretation specifies that two analogies be applied in order to provide a picture of what happens to a quantum system between observations. The two analogies are to moving particles and spreading waves.

These two analogies are recognized to be mutually exclusive. A particle is localizable; a pure wave is not localizable. Only one of the two analogies is applicable to the results achieved within a specific experimental arrangement.

Consider the case of a beam of electrons passing through a slit in a diaphragm so as to impact a photographic plate on the other side. If the diaphragm is rigidly mounted, the appropriate analogy is to a wave passing through and spreading out from the slit. If the diaphragm is spring-mounted, so as to be free to move in the vertical plane, the appropriate analogy is to particles exchanging momentum and energy with the diaphragm.

Bohr indicated that the wave picture provides a spatio-temporal description of the diffraction process, and that the particle picture provides a causal analysis—*qua* energy/momentum transfer—of the process. He stressed that, for a particular experimental arrangement, one may provide either a spatio-temporal description or a causal analysis of the process, but not both.

On the Copenhagen interpretation, the use of two mutually exclusive pictures is complementary. A complete interpretation of a quantum process requires application of both pictures. The Copenhagen interpretation of quantum theory fits the hypothesis-dictionary-analogy format, given the use of mutually exclusive, but complementary, analogies.

Critics of the Copenhagen interpretation have emphasized that the hypothesis-plus-dictionary of quantum theory suffices for the prediction of experimental results in the quantum domain. The analogy to classical waves and particles is irrelevant. The Copenhagen interpretation may fit the hypothesis-dictionary-analogy model of a "mechanical theory." However, there are many physicists who reject as superfluous the addition of analogical wave and particle pictures to the hypothesis-plus-dictionary.

Mathematical Theories and Analogies

Campbell maintained that a successful "mathematical theory" also displays an analogy. In the case of Fourier's theory of heat conduction, the analogy is "based on the laws to be explained."[9] The analogy is one of form. The laws deducible from the theory share a common mathematical form with the hypothesis of the theory.

Fourier's theory can be applied to specify rates of heat conduction in finite slabs of various materials. One substitutes values of density, thermal conductivity, and specific heat for the material in question. The result is an empirical law appropriate to that material. Empirical laws obtained in this way exhibit the form of the hypothesis from which they are deduced.[10]

Campbell insisted that Fourier's theory was not a mere shorthand representation of the empirical laws derivable from it. Alternative theories can be invented to imply the same empirical laws. Campbell's artificial theory of the temperature dependence of the electrical resistance of metals was designed to illustrate this point.

Carnap on Elementary Terms and Abstract Terms

Rudolf Carnap restated the hypothesis-plus-dictionary view of scientific theories in an influential essay published in the *International Encyclopedia of Unified Science* in 1939. He declared that

> any physical theory, and likewise the whole of physics, can . . . be presented in the form of an interpreted system, consisting of a specific calculus (axiom system) and a system of semantical rules for its interpretation.[11]

This claim was repeated by Philipp Franck and Carl Hempel in subsequent essays in the same encyclopedia.[12]

The originality of Carnap's presentation was a distinction he drew between "elementary terms" and "abstract terms." He suggested that terms can be graded, at least roughly, with respect to abstractness.

degree of abstractness

──▶

"red"   "length"   "character"   "velocity of a molecule"
"smooth"

"Red" and "smooth" are "elementary terms" because the truth-value of "that object is red (smooth)" can be determined directly by observation. "Length" is a bit more abstract, since the assignment of length to an object typically requires establishing coincidences with a properly calibrated rod. "Character" in Medelian genetics is still more abstract, since statements about a character are correlated with "direct observation" only indirectly *via* dictionary entries to phenotypic properties. The "velocity of a molecule" in the kinetic theory of gases is more abstract yet, since it is correlated with observations only through its contribution to the root-mean-square velocity of an ensemble of molecules, which root-mean-square velocity is correlated, in turn, with the relatively elementary terms "pressure" and "temperature."

On the assumption that distinctions of this kind can be drawn, Carnap posed two questions about the structure of a theory whose axiom system contains both elementary terms and abstract terms. The first question was "which terms should have dictionary entries, elementary terms or abstract terms?" Carnap noted that it is possible to assign dictionary entries to any of the terms of the axiom system. It is possible to interpret an axiom system by providing semantical rules only for the most abstract terms in the axiom system. However, if one's purpose is to facilitate applications of the theory, then semantical rules should be given instead for the elementary terms. Carnap maintained that this purpose should guide the theorist. From this standpoint

> a rule like "the sign 'P' designates the property of being blue" will do for the purpose indicated; but a rule like "the sign 'Q' designates the property of being electrically charged" will not do.[13]

The second question about theory-construction Carnap posed was "which terms should be primitives in the development of the axiom system?" He outlined two methods for the development of an axiom system:[14]

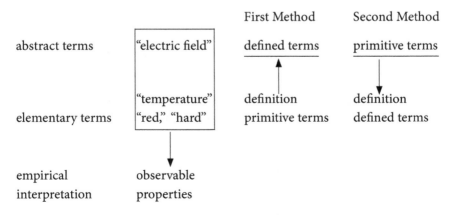

In the first method, elementary terms are primitives, and abstract terms are introduced by definitions from the elementary terms. But since this introduction of abstract terms can be accomplished neither by explicit definitions nor by contextual definitions, the first method is impractical (see pp. 13–14). One cannot start with "directly observable" properties and generate from them, by any straightforward procedure, the abstract terms of scientific theories. It is for this reason that scientists follow the second method.

In this second method, the abstract terms of the axiom system are primitives and the "more elementary" terms are introduced by definitions from the abstract terms. Carnap challenged theorists to formulate axiom systems by beginning with abstract terms and working downward toward directly observable property terms. He suggested that ideally this formulation would be achieved by utilizing explicit definitions alone. But he was quick to point out that there are formidable difficulties for such a program. Scientists utilize many terms for which no definitions are available by reference to "more abstract" terms.

Carnap likened theory-construction to the spinning of a web. However, it is an odd sort of spinning, for the web is not attached to anything until the very end. Carnap declared that

> the calculus is first constructed floating in the air, so to speak; the construction begins at the top and then adds lower and lower levels. Finally, by the semantical rules, the lowest level is anchored at the solid ground of the observable facts.[15]

Each horizontal strand of the web relates terms at a given level of abstractness. Progression down the web upon application of method (2) is movement from the abstract to the concrete. The lowest strand of the web is anchored to the plane of observation by means of semantical rules.

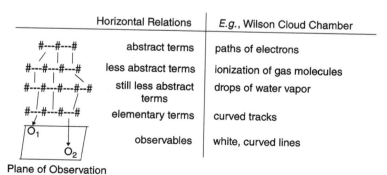

	Horizontal Relations	*E.g.*, Wilson Cloud Chamber
#---#---#	abstract terms	paths of electrons
#---#---#---#	less abstract terms	ionization of gas molecules
#---#---#---#--#	still less abstract terms	drops of water vapor
#---#---#---#	elementary terms	curved tracks
O_1 ... O_2	observables	white, curved lines

Plane of Observation

Figure 13 Carnap's "Spider-Web" Image of Theories.

Carnap noted that, on method (2), abstract terms are taken as primitives and semantical rules are given only for elementary terms. For these reasons, semantical rules provide only a partial interpretation for the axioms. Fortunately this suffices for the purposes of description and prediction. It is only singular sentences about elementary terms that can be tested directly. Semantical rules link such sentences to sentences about the directly observable properties of bodies, thereby converting the axiom system into an empirically significant theory.

Hempel's "Safety-Net" Image

Vertical relationships among levels that differ in abstractness were a prominent feature of Carnap's "spider-web" image of scientific theories. In Hempel's version of the hypothesis-plus-dictionary view of theories, the only vertical relationships are the semantical rules that link terms of the axiom system to the plane of observation. It is as if the entire system floats in a single horizontal plane above the plane of observation. The resultant image bears some resemblance to safety-nets used for the protection of trapeze artists.

The net (axiom system) is supported by poles (semantical rules) which rest on the arena floor (observation reports). Hempel agreed with Carnap that the axiom system of a scientific theory must touch base on the plane of observation reports. He wrote that

> science is ultimately intended to systematize the data of our experience, and this is possible only if scientific principles, even when couched in the most esoteric terms, have a bearing upon, and thus are conceptually concerned with, statements reporting in "experiential terms" available in everyday language what has been established by immediate observation.[16]

Campbell had emphasized that it is not necessary that there be a semantical rule for each term of an axiom system. The net does not require support at every knot. But there must be some support, otherwise the net will collapse. How many supporting poles (semantical rules) are necessary if the net is to fulfill its role?

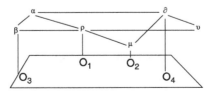

Figure 14 Hempel's "Safety-Net" Image of Theories.

Hempel suggested that we could answer this question if there were available a satisfactory theory of quantitative confirmation. Such a theory would stipulate the degree of evidential support conferred upon each theorem of a scientific theory by each observation report. We then could correlate the structure of a theory with its degree of confirmation, and draw an informed conclusion about the support provided for its net. Hempel conceded, however, that no confirmation theory then available was adequate for this task.[17] His suggestion to measure the bond between axiom system and observation reports by a theory of confirmation had the status of a program for future inquiry.

Braithwaite's Formalist Interpretation of Theories

R. B. Braithwaite presented a formalist interpretation of scientific theories in *Scientific Explanation* (1953).[18] In agreement with Campbell, Carnap, and Hempel, Braithwaite held that a scientific theory is an empirically interpreted calculus (i.e., axiom system).

where $\alpha, \beta, \gamma, \partial$ are primitive elements
            t_1, t_2, t_3, t_4 are theoretical terms
            #1, #2, #3 are derived formulas
            i, ii, iii are theorems

According to Braithwaite, a calculus "represents" a deductive system provided that:

1. each theoretical term corresponds to a primitive element;
2. the axioms and the initial formulas are isomorphic; and,
3. to each rule of inference there is a corresponding rule of symbolic manipulation.

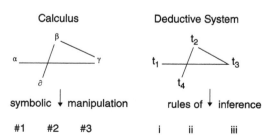

Figure 15 Deductive Systems as Interpreted Calculi.

Calculus	Deductive System
$d^2x / d y^2 = k$	$d^2s / d t^2 = g$
|	|
Rules of Integral Calculus	Rules of Integral Calculus
↓	↓
$x = 1/2 k y^2 + A y + B$	$s = 1/2 g t^2 + v_0 + s_0,$
	and for $v_0 = 0$ and $t_0 = 0,$
	$s = 1/2g t^2$

Figure 16 Galileo's Theory of Free-Fall.

These conditions are fulfilled by the reconstruction of Galileo's theory about freely falling bodies presented in Figure 16.

where x and y are variables
where s is the distance of fall in time t
         and k, A, and B are constants

Galileo's deductive system receives an empirical interpretation when values of the symbols s and t are correlated with the results of concomitant measurements of distance and time.

An important portion of thermodynamics also may be set forth as a calculus that represents a deductive system (see Figure 17).

Relations within the deductive system are given empirical significance upon correlation of values of P, V, and T with experimentally determined values of pressure, volume, and temperature.

Equation (4) then relates the compressibility of a gas to its rate of thermal expansion and rate of pressure increase with temperature. This relation has been confirmed experimentally for a variety of cases over fairly extensive ranges of pressure and temperature.

Braithwaite drew upon this analysis of the structure of scientific theories in a discussion of two important methodological problems. The first problem is to explain how theoretical terms acquire empirical significance.

The Empirical Significance of Theoretical Terms

In the case of the two theories discussed above—Galileo's theory of free-fall and the thermodynamic behavior of gases—each non-logical term of the deductive system is correlated with an experimentally measurable property of physical systems. However, the empirical interpretation of deductive systems often is more complex. As Campbell had emphasized, many scientific theories contain terms that are not correlated directly with observable properties. Examples of

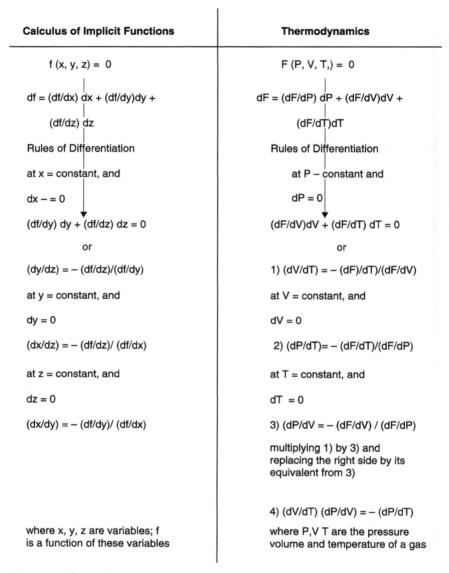

Calculus of Implicit Functions	Thermodynamics
f (x, y, z) = 0	F (P, V, T,) = 0
df = (df/dx) dx + (df/dy)dy +	dF = (dF/dP) dP + (dF/dV)dV +
(df/dz) dz	(dF/dT)dT
Rules of Differentiation	Rules of Differentiation
at x = constant, and	at P – constant and
dx – = 0	dP = 0
(df/dy) dy + (df/dz) dz = 0	(dF/dV)dV + (dF/dT) dT = 0
or	or
(dy/dz) = – (df/dz)/(df/dy)	1) (dV/dT) = – (dF)/dT)/(dF/dV)
at y = constant, and	at V = constant, and
dy = 0	dV = 0
(dx/dz) = – (df/dz)/ (df/dx)	2) (dP/dT)= – (dF/dT)/(dF/dP)
at z = constant, and	at T = constant, and
dz = 0	dT = 0
(dx/dy) = – (df/dy)/ (df/dx)	3) (dP/dV = – (dF/dV) / (dF/dP)
	multiplying 1) by 3) and replacing the right side by its equivalent from 3)
	4) (dV/dT) (dP/dV) = – (dP/dT)
where x, y, z are variables; f is a function of these variables	where P,V T are the pressure volume and temperature of a gas

Figure 17 Thermodynamics as a Deductive System.

such terms are "individual molecular velocity" in the kinetic theory of gases and "character" in Mendelian genetics. Within these two theories no empirical procedures are specified to determine values of these concepts. Nevertheless, the axioms that make assertions about these theoretical terms are believed to be empirically significant.

To explain how such terms acquire empirical meaning Braithwaite suggested an analogy to the operation of a zipper.[19] A zipper is effective only if a tug on its tab engages the two sides so as to transfer upwards the security of the anchor point. In similar fashion, empirical significance is conferred upwards in a deductive system from theorems to axioms. This conferring of meaning is accomplished by virtue of an isomorphism of structure of the deductive system and the calculus.

Braithwaite claimed that the appropriate answer to a question about the meaning of a theoretical term is:

1. to exhibit the structure of the deductive system by showing how it represents a calculus (viz., to show how the teeth engage), and
2. to indicate how the theorems of the deductive system receive empirical interpretation (viz., to show how the zipper is anchored at its base).

For instance, what the 'Ψ-function" means is just the syntax of its use in the deductive structure of quantum theory, together with the interpretation of the theorems of the theory as laws about electron charge densities, scattering distributions, and the like.

Braithwaite insisted that it is more fruitful to discuss the meaning of theoretical terms than it is to discuss the "reality" of the entities that theoretical terms supposedly denote. He suggested that if a person persists in asking questions about the "reality" of the theoretical entities denoted by such terms as "electron," "entropy," and "Ψ", then the following answer is proper: to claim that a theoretical entity "exists" is to claim nothing more than that the corresponding theoretical term occurs in a true theory.[20]

The Status of Models

A second problem addressed by Braithwaite is the status of models. Braithwaite acknowledged that scientists and philosophers use the term "model" in diverse ways. One widespread view is that a "model" is a picture of some underlying mechanism that is responsible for the processes addressed by the theory.

Braithwaite restricted use of the term "model" to deductively organized groups of propositions that satisfy certain logical requirements. He held that, given a deductive system that interprets a calculus, any alternative interpretation of that calculus is a model of the deductive system.[21] In order to qualify as an alternative interpretation of a calculus, the model must be isomorphic in structure to the deductive system. To each term of the deductive system there must be a corresponding property term of the model. And the terms within the

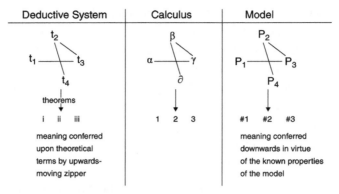

Figure 18 Braithwaite on Scientific Models.

deductive system and the property terms of the model must be related in the same manner, as shown in Figure 18.

Braithwaite admitted that a model can be of heuristic value if a consideration of its properties suggests extensions of the corresponding deductive system. Sometimes this is a matter of noting relations other than those stated in the model, and then postulating corresponding relations to hold among the primitives of the deductive system. Or perhaps there is some further property of the model—P_5—which suggests addition of a term t_5 to the deductive system. For instance, consideration of the billiard-ball model for the kinetic theory of gases may suggest modifying the axioms of the theory to include assertions about molecules of finite diameter. And consideration of the planetary model for the Bohr theory of the hydrogen atom may suggest its extension to electronic orbits of elliptical form. In these cases, the initial theory is replaced by a theory of greater complexity.

A model, then, can provide hints for replacing one theory by another. Whether such hints will prove fruitful cannot be stated in advance. Braithwaite cautioned that, in many cases, preoccupation with models has proved stultifying. In a case where the *only* thing which a model and a deductive system have in common is similarity of structure, it is futile to develop new deductive systems by analogical extrapolation from additional properties of the model.

On Braithwaite's view of models, it is not necessary to understand the model in order to understand the corresponding theory. In the deductive system, axioms acquire empirical meaning from below by an upward movement of the zipper. The axioms are logically prior, but epistemologically posterior, to the theorems. In the model, by contrast, the initial propositions are given a direct empirical interpretation independently of any interpretation of propositions derived from them. The initial propositions about the properties of the model are both logically and epistemologically prior to the derived propositions.

For this reason, reference to a model may help one to see what are the deductive relationships that hold within a theory. But every deductive relationship within the model has its counterpart in the deductive system. Consequently, there is no reason to include any of the propositions of the model in the premises of scientific explanations. A model may be a useful heuristic device, but it is not an essential component of a scientific theory.

Braithwaite's position is opposed to the position previously staked out by Campbell. Campbell had maintained that a scientific theory explains laws deducible from it only if it displays an analogy to other known laws. He supported this position by creating an artificial theory that implies the law of the temperature dependence of the electrical resistance of pure metals. According to Campbell, this artificial theory fails to explain the law because it lacks an analogy to previously known laws.

Hempel supported Braithwaite's position on this matter. Referring to Campbell's artificial theory, he derived the law of temperature dependence of the electrical resistance of pure metals from a different theory that does display an analogy to known laws. Hempel's theory is shown in Figure 19.

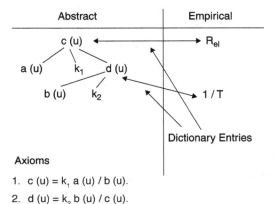

Figure 19 Hempel's Theory of the Temperature Dependence of the Electrical Resistance of Pure Metals.

Source: Carl Hempel, *Aspects of Scientific Explanation* (New York: Free Press, 1965), 444–5.

The hypothesis implies that c (u) = $k_1 k_2$ d (u). This relation, interpreted by reference to the dictionary, states that the electrical resistance of a piece of pure metal is directly proportional to its absolute temperature. This is the same law that Campbell had derived from his artificial theory.

However, Hempel's theory, unlike Campbell's theory, does display an analogy to a previously established law. Each of the two axioms of Hempel's hypothesis has the same mathematical form as Ohm's Law—$i = V / R$—where i is the current in a circuit, V is its voltage, and R is its resistance.

This formal analogy provides no support for the theory. Hempel insisted that the fact that a theory displays an analogy to previously established laws does not, in itself, create explanatory value for a theory.

Hempel surely is correct. However, Campbell, if he were made aware of this objection, likely would respond that a successful theory must display *some* analogy to previously lawful regularities. The fact that a particular analogy fails does not invalidate this requirement.

Margenau's Constructionist Philosophy of Science

Henry Margenau combined a phenomenalist interpretation of the hypothesis-plus-dictionary view of theories with a Kantian emphasis on regulative principles that prescribe the form of acceptable theories. The offspring of this marriage was a "constructivist" philosophy of science.

Margenau noted that certain elements of experience are characterized by spontaneity, passivity, irreducibility, and relative independence. These elements of experience include both sense impressions and dreams, hallucinations, and illusions. At the level of spontaneity and passivity, waking impressions and dreams are on a par. However, Margenau believed that a subdivision could be established within experience. Sense impressions, but not dreams and hallucinations, can be linked to constructs that satisfy certain formal and empirical requirements.

The first step in the transition from the data of immediate experience to scientific knowledge is the selection of "rules of correspondence." In Margenau's usage, "rules of correspondence" link impressions to constructs. A variety of types of rules of correspondence are utilized in the formation of scientific theories. One type associates sense data with an object that is posited as its source; "external object" is a construct posited to account for regularities among our sense impressions. More abstract rules of correspondence correlate "wavelength" with color and "electron path" with tracks observed in a Wilson Cloud Chamber.

The second step in the transition to scientific knowledge is the conversion of constructs into "verifacts." One condition that a construct must satisfy in order to qualify as a verifact is participation in circuits of verification (see Figure 20).

A circuit of verification begins and ends in the data of immediate experience. A transition from the perceptually given (P_1) to a construct (C_1) is accomplished by selection of a rule of correspondence. If C_1 has a use in science, it will be related formally to other concepts through axioms, theorems, and definitions. In virtue of these formal relations, it may be possible to recross the boundary between the abstract and the empirical by selecting a rule of correspondence that terminates at P_2.

If P_2 corresponds to a datum of immediate experience, then the network of concepts traversed in the circuit has been confirmed. Margenau maintained that a necessary condition of the "validity" of a theory, and the "verifact-status" of its constructs, is that it participates in a "sufficient number" of circuits of verification. Margenau thus reformulated the hypothesis-plus-dictionary view of theories on a phenomenalist basis. Dictionary entries link the terms of scientific theories to the data of immediate experience.

One might wonder how this circuit of confirmation can get started. Margenau noted that sensations on the P-plane are subjective. There is no objective standard against which they may be compared. I cannot judge whether your sensation in a given context is the same as mine. P-plane sensations also are variable. The intensity of a type of sensation depends on the physiological state of a given subject. Experienced "forces," "times," and "temperatures" are an insufficient basis for science. These sensations are of value within science only when replaced by

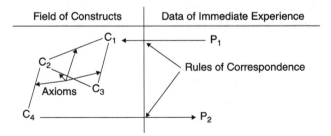

Figure 20 Margenau's Circuit of Verification.

Source: Henry Margenau, *The Nature of Physical Reality* (New York: McGraw-Hill, 1950), 102–7.

constructs that are intersubjective, stable, and quantitative. Margenau observed that

> the equations of mechanics do not contain the pressure on my hand, nor the sense of heat in my fingertips, nor the flight of consciousness that amounts to a passage of time. Science converts each of these into an idea that *corresponds* to it.[22]

Margenau provided illustrations of how this conversion takes place. The pressure on one's hand can be correlated with, and replaced by, dynamometer readings, the sense of heat by thermometer readings, and the flight of consciousness by clock readings. The replacement readings satisfy the conditions of intersubjectivity, stability, and quantitative variation. Margenau maintained that the conversion process is in each case governed by a "rule of correspondence." He claimed that, in the cases cited, the rules of correspondence are what Bridgman called "operational definitions."[23]

This surely is incorrect. Bridgman's operational definitions introduce a concept into science by the formula "if an operation of a certain type is performed, then the concept has a specified value if, and only if, specified results of the operation are observed." For instance, "if a substance is placed in a liquid, then it is soluble if, and only if, it dissolves." However, Margenau specified no operations for introducing "force" by reference to results of perceived pressure upon one's hand. Nor did he specify operational definitions, in Bridgman's sense, to link "time" or "temperature" to the appropriate sensations.

Nevertheless, Margenau declared that

> passage from the feeling in my fingertips to the recording of the thermometer is regulated, or is performed, via a rule of correspondence.[24]

This is unfortunate linguistic usage. Rules of correspondence link terms of an axiom system—e.g., individual molecular velocities in the kinetic theory of gases—with observables—e.g., "temperature" and "pressure." Clearly, Margenau is using "rule of correspondence" in an extended sense, not restricted to the connection between terms of an axiom system and observables.

How does one use this expanded sense of "rule of correspondence" to establish a transition from perceived to objective senses of force, time, and temperature? Margenau discussed various periodic processes that may be employed to measure intervals of time—the motions of pendulums, the sun, the stars, the vibrations of quartz crystals, and the vibrations of atoms or nuclei.[25] He noted that scientists select a best procedure to measure time by showing which periodic processes introduce greatest regularity in data that confirm basic laws of motion.

Margenau declared that

> clearly, then, a choice is made in the application of a rule of correspondence, the choice being that which leads to the simplest ... law of nature.[26]

Margenau applied the term "rule of correspondence" to the choice of a most effective measure of time for the purposes of science.

However, the "rule of correspondence" that guides selection of the best measure of time does not connect perceived "flights of consciousness" with an intersubjective construct of time. The question remains, what has the use of a "rule of correspondence" in this extended sense to do with instances of perceived duration?

Margenau appears to be using "rule" in a Wittgensteinian sense. Wittgenstein emphasized the intersubjective character of "obeying a rule."[27] One cannot obey a rule privately. To think that one is obeying a rule does not establish that one is obeying a rule. Rules are intersubjectively accessible. If a rule exists, then it may be obeyed interchangeably by a variety of participants.

Wittgenstein insisted that to obey a rule is to engage in a "form of life." What determines whether a rule is followed in a particular instance depends on the *actions* of the subject (and not just his testimony), and the reactions of other persons to what he does.

It is not always clear whether specific behavior qualifies as implementation of a particular rule. Suppose a subject writes the following sequence of numbers: 1, 1, 2, 3, 5, 8. An observer of this behavior might be incorrect to conclude that the subject was reproducing the Fibonacci series and that the next term in the sequence is 13. The subject might have been beginning a repeating sequence of numbers—1, 1, 2, 3, 5, 8, 1, 1, 2, 3, 5, 8, ... There is no basis on which to judge which rule is being followed over and above individual and communal practice.

Scientists regularly and consistently make transitions from perceptual awareness to constructs. They replace subjective sensations of force, time, and temperature with constructs that are intersubjectively applicable, reliable, and quantitative. This is sufficient to qualify such behavior as "rule-governed," and to inaugurate transitions from P-plane data to constructs *via* "rules of correspondence." A "rule of correspondence," in Margenau's usage, is whatever establishes connections between "sensed nature" and constructs.[28]

In the essay "Metaphysical Elements in Physics" (1941), Margenau distinguished his starting-point from that of "orthodox" theorists such as

Carnap and Hempel. Whereas Carnap and Hempel begin the discussion of confirmation on the "secondary plane of language," i.e., with observation reports, Margenau resolved to begin his analysis of circuits of empirical confirmation with the "primary elements of experience," elements of "sensed nature."[29] It is a "fundamental error," he declared, to confuse "primary scientific experience with language."[30]

Hempel was critical of the attempt to link the analysis of scientific theories to a language that records the data of immediate experience. He pointed out that there are two difficulties for a phenomenalist grounding of scientific theories. In the first place, there is no linguistic framework for the use of phenomenalist terms. And in the second place, restricting confirming instances to "color images," "tactile sensations," and the like deprives scientific theories of an intersubjective test-basis.

Since scientists do advance consistently from perceptual experience to constructs, they do follow "rules of correspondence" (in the above sense). Margenau declared that

> these rules are sometimes intuitive and unstated, more often operational procedures, occasionally of a more intricate logical character. If necessary, they *can* always be stated, and alas, they can be stated in a vast variety of ways.[31]

It is an "intuitive and ingrained" rule of correspondence that governs the transition between experienced patches of light and the construct "desk." An "operational" rule of correspondence then governs the transition between these experienced patches of light and the construct "mass."[32]

Given a set of basic constructs obtained by applying rules of correspondence to data on the P-plane, circuits of empirical confirmation begin and end with measurements made of the values of these basic constructs. These measurements can be performed, in principle, by any qualified observer.

Margenau maintained that participation in circuits of verification is a necessary but not a sufficient condition to establish a construct as a verifact. In addition the construct must be embedded in a theory that conforms to regulative principles that stipulate the "proper form" of organized scientific knowledge. These principles are:

1. The construct must be multiply connected by formal relations to other constructs.
2. The construct must retain uniformity of meaning in all its applications within the theory.

3. The construct is embedded in a theory from which laws of causal form may
   be derived. Margenau held that a causal law specifies a determinate
   unfolding of the states of a physical system. From a causal law, together with
   information about the state of a system at one time, one can deduce
   information about the state of the system at future, or past, times. The
   mathematical form of a causal law is a differential equation that does not
   contain the time variable in explicit form.

Galileo's law of free-fall is an example of a law of causal form. From this
law, together with information about the position and velocity of a body at one
instant of time, one can predict the subsequent positions and velocities of
the body.

4. The theory should be "simple" and "elegant."
5. The construct should be capable of incorporation into additional theories.[33]

Margenau identified four constituents in scientific knowledge—the data of
immediate experience, rules of correspondence, constructs, and regulative
principles. Science progresses from data to constructs *via* rules of correspondence.
When this construction produces verifacts, the goal of science has been achieved.
Moreover, reality has been created. Margenau proposed as a criterion of reality
that, in all cases, if x is a verifact, then x is real. Insofar as a construct functions in
well-confirmed laws of causal form, it is part of reality. From this point of view,
the Ψ-function is as real as Mount Everest. And the real world comprises all
verifacts and those data of experience that are correlated with verifacts by rules
of correspondence.[34]

Margenau's version of the hypothesis-plus-dictionary view of theories
represented a marked shift of emphasis. Carnap and Hempel had sought to
anchor scientific theories in the bedrock of a physicalistic language. Their
common assumption was that the truth or falsity of statements that assign values
to the "directly observable" properties of bodies may be ascertained independently
of the truth or falsity of theoretical claims. They believed that the existence of a
theory-independent observation language is necessary both to confer empirical
significance on theoretical terms and to test theories in which these terms occur.
From this standpoint, the important question about theoretical terms is how
statements about such terms can be related to observation reports.

Margenau, by contrast, did not accept the orthodox position on the existence
of a theory-independent observation language. Instead he gave a Kantian turn to
the question about theoretical terms. For Margenau the important question is

"given the possibility that some of our constructs are verifacts, what are the necessary and sufficient conditions of this possibility?" The answer he gave placed emphasis on empirical confirmation and regulative principles that stipulate conditions under which scientific knowledge is achieved.

The Realist–Instrumentalist Dispute

During the immediate post-war period, realism and instrumentalism were the most frequently held positions on the cognitive status of scientific theories. The realist position is that theories are statements that function as premises in important scientific explanations. On the realist view, questions about the truth of a theory are relevant to an assessment of the adequacy of the explanations in which it figures.

The instrumentalist position is that a theory is an instrument for the derivation of statements about observations. A theory provides a schema for the replacement of statements about initial conditions and boundary conditions with further statements about observable properties and relations. On the instrumentalist view, questions about the truth of a theory do not arise.

Post-war orthodox theorists were concerned primarily with applications of logic to the problems of confirmation and explanation. The cognitive status of theories was not the burning issue. Those who did take a position on this issue often selected the instrumentalist alternative. Philipp Frank expressed the instrumentalist position as follows:

> this metaphysical concept of a true theory as a "replica of physical reality" is not prevalent in the scientific philosophy of today. A theory is now rather regarded as an instrument that serves some definite purpose. It must be helpful in predicting future observable facts on the basis of facts that have been observed in the past and in the present. The theory should also be helpful in the contribution of devices which can save time and labor.[35]

Frank gave credit to Heinrich Hertz for developing a "positivist" view of scientific theories.[36] Many nineteenth-century scientists had sought in vain for entities denoted by the terms that occur in Maxwell's equations for the electromagnetic field. Hertz concluded that this search was misguided. He recommended that attention be shifted away from the question "What is described by the terms of Maxwell's equations?" According to Hertz, it is more fruitful to ask "How does electromagnetic theory enable the description and prediction of phenomena?"[37]

If a theory is successful in enabling the prediction of phenomena, it is because the interpreted axiom system entails experimental laws. In the 1930s, Frank Ramsey and Moritz Schlick suggested that experimental laws themselves are material rules of inference.[38] On their view, scientific inference proceeds from observation statements to observation statements according to patterns stipulated by laws. For example, the fact that a piece of blue litmus paper turned red can be deduced from premises that state the presence of blue litmus dye on the paper and the immersion of the paper in an acidic solution, viz.,

$$\frac{\text{Ba \& Aa}}{\therefore \text{Ra}}$$

The experimental law that all blue litmus turns red in acid is not a statement that can be true or false. It is, rather, a schema for drawing inferences, viz.,

> any statement of the form "(Bx & Ax)" may be replaced by a statement of the form "Rx."

This instrumentalist interpretation of laws was reaffirmed in the 1950s by Gilbert Ryle and Stephen Toulmin. Ryle remarked that a scientific law is used as an "inference ticket." The law

> licenses its possessors to move from asserting factual statements to asserting other factual statements.[39]

Toulmin supported Ryle's position. In addition, he emphasized that, although laws themselves are neither true nor false, statements about their ranges of application are.[40] On Toulmin's view, it is true to say that Boyle's Law does not hold at high pressures. At high pressures, inferences drawn from initial conditions using Boyle's Law fail to agree with experimental results. It would be a mistake, however, to say that Boyle's Law itself is false. The law is merely a schema for drawing inferences. It is useful in certain applications and not useful in other applications.

Philosophers of science who place an instrumentalist interpretation on experimental laws also take theories to be conceptual devices for the derivation of observation statements. An interesting variant of the instrumentalist view is the position of "eliminative fictionalism." The eliminative fictionalist concedes that references to statements about theoretical terms is useful. However he holds that theories are "useful" in the same way that the notation x^6 is useful. The exponent 6 is eliminable in favor of the product of six terms. Similarly, statements about theoretical terms are eliminable in terms of statements about observation terms.

The position of eliminative fictionalism received some support from a theorem proved by William Craig in 1953.[41] Craig showed that, granted certain widely-fulfilled conditions, for every system (1) which contains both theoretical terms and observation terms, there is a system (2) which contains only observation terms, such that each observationally significant theorem of system (1) also is a theorem of system (2). Thus, reference to theoretical entities can be eliminated from scientific theories without sacrificing empirical content.

There is a price to be paid, however. Hempel emphasized that the replacement system (2) contains infinitely many postulates. System (2) may have the advantage of being formulated exclusively in statements about observation terms. But it is unmanageable in any practical sense. System (2) lacks functional utility.[42]

Nagel's Analysis of the Realist–Instrumentalist Dispute

In *The Structure of Science* (1951), Ernest Nagel sought to dissociate orthodoxy, whose logical reconstructionist orientation he shared, from a commitment to instrumentalism.[43] He conceded that instrumentalists are correct to emphasize the role of limiting concepts and theoretical concepts in scientific theories. Scientists do use phrases such as "bodies subject to no impressed forces," "point-particles," and "light waves" in theories that are used to predict experimental results.

Realists maintain that certain generalizations about the "entities" designated by such terms are factually true. But if an empirically meaningful statement is true only if it asserts some relation between existing things or events, or their properties, then it is difficult to see how the realist's claim can be made good. If the law of inertia can be true only if there exist force-free bodies, and is false if there are no such bodies, then this law is false. And since realists claim that theories are premises in explanatory arguments, no argument whose premises include Newton's law of inertia is sound. Were this the case, we would be hard pressed to justify the rationality of the actions that we base on the conclusions of such arguments.

This objection to realism is not decisive. The realist need not insist that each statement within a theory is empirically significant in isolation from the remainder of the theory. Indeed, the realist may adopt Braithwaite's "zipper image" to explain how theoretical terms acquire empirical meaning from below. If the axiom system as a whole is linked to experience by semantical rules, then the theory may be used as a premise in explanatory arguments.

	Context #1	Context #2
Rules of	A statement of the form	(x)(Px ⊃ Qx); (p⊃q) & p
Inference	'x expands' is deducible from	∴Pa⊃Qa ∴q
Employed	a statement of the form	
	'x is copper and is heated'	
Argument	Ca & Ha	(x) [(Cx & Hx) ⊃ Ex]
	—————	
	∴ Ea	Ca & Ha
		—————
		∴ Ea

Figure 21 Nagel on Material Rules of Inference and the Deductive-Nomological Pattern.

Hence it is not inconsistent to maintain that a theory is both instrumentally useful and true. Nagel pointed out that one and the same law may serve in one context as a rule of inference and in a second context as a premise (see Figure 21).[44]

The realist prefers context (2), in which the law appears as a premise in an argument whose conclusion is deduced by use of the formal rules "universal instantiation" and "*modus ponens*." Suppose that a critic complains that she cannot see any connection between the premises and the conclusion of the argument in context (2). The standard defense is to exhibit the *modus ponens* form—[(p⊃q) & p /∴ q]—and show that the argument in context (2) has the same form. If the critic protests that she sees no reason to accept *modus ponens* as a valid argument form, then she has opted out of the game in which "argument" and "deductive validity" are used. Given the truth-table definitions of "⊃" and "&," and the meaning of "deductive validity," the *modus ponens* form qualifies as valid. Of course the critic is free to reject the entire system of deductive logic in which these meanings are assigned. But then, it is incumbent upon her to indicate the rules of inference that she does take to be valid.

The instrumentalist prefers context (1), in which the law itself is used as a rule of inference. Suppose a critic complains that he cannot see any connection between the premise and the conclusion of the argument in context (1). Why, he asks, should one accept that "*x* expands" is deducible from "*x* is copper and *x* is heated"? This objection is different in kind from the objection to *modus ponens*. The critic has not opted out of the game in which "argument" and "deductive validity" are used by asking why a particular material rule of inference ought to be accepted.

It would seem that whatever evidence counts for the usefulness of this rule of inference also counts as confirming evidence for the corresponding law. But if grounds for accepting a material rule of inference also is confirming evidence

for the corresponding law, why is it important to take sides in the instrumentalist–
realist dispute?

Nagel's answer was that there is only a verbal difference between asserting
"*L* is satisfactory as a rule of inference" and "*L* is a true premise in explanatory
arguments." Instrumentalists and realists agree on how to conduct scientific
inquiry, and they agree on the important results that have been achieved. Both
sides acknowledge the power of relativity theory, quantum mechanics, and the
theory of organic evolution. Nagel concluded that the instrumentalist–realist
dispute is a dispute over "preferred modes of speech." Instrumentalists and realists
express different preferences for accommodating language to the generally
admitted facts. [45]

Reduction and the Growth of Science

Nagel's Conditions for Successful Reduction

Science has a history as well as a logical structure. Orthodox theorists focused
primarily on structure. With respect to history, there was general agreement that
there had been progress in science. Successive theories have provided increasingly
more precise and extensive descriptions of the world.

Ernest Nagel sought to uncover the conditions responsible for this success.
He introduced the notion of the "reduction" of one theory to another. Nagel
declared that

> the phenomenon of a relatively autonomous theory becoming absorbed by, or
> *reduced* to, some other more inclusive theory is an undeniable and recurrent
> feature of the history of modern science.[1] [emphasis mine]

Given a sequence of theories—*P*, *Q*, . . .—prior theory *P* is reduced to successor
theory *Q* provided that *Q* incorporates *P* such that certain conditions are fulfilled.

Nagel distinguished two types of reduction. In a "homogeneous reduction,"
the concepts of *Q* are "substantially the same as" the concepts of *P*.[2] Examples
include the replacement of Galileo's theory of freely falling bodies by Newton's
theory of mechanics, and the replacement of Bohr's theory of the hydrogen atom
by Sommerfeld's theory that added elliptical electron orbits.

"Heterogeneous reductions" are more interesting. In a heterogeneous reduction,
Q lacks some of the concepts in which *P* is expressed. Nagel cited the transition
from classical thermodynamics to kinetic molecular theory as an instance of
heterogeneous reduction. The kinetic theory is a statistical theory about the

motions of molecules. Classical thermodynamics is stated in the concepts of "temperature," "pressure," and "volume."

Nagel set forth the following conditions for the successful reduction of *P* to *Q*.

Formal Conditions

1. Connectability. There is a connecting statement that links each term of *P* with the terms of *Q*.
2. Derivability. The empirical laws of *P* are derivable from theory *Q*.

Empirical Conditions

3. Evidential support. There is evidence that supports *Q* over and above the evidence that supports *P*.
4. Fertility. The theoretical assumptions of *Q* give rise to the further development of *P*.[3]

Nagel maintained that the transition from thermodynamics to the kinetic theory of gases satisfies the above conditions.[4] The terms "temperature" and "pressure" are linked to the "root-mean-square velocity" of molecules (μ) by the relations "$T = n\ m\ \mu^2/\ 2k$" and "$P = n\ m\ \mu^2/\ V$."

The laws of Boyle, Charles, and Gay-Lussac are deductive consequences of the axioms of the kinetic theory, given the connecting links to "temperature" and "pressure." There is extensive evidence from domains other than thermodynamics that supports the kinetic theory.

The transition from thermodynamics to kinetic theory proved fertile as well. The laws of Boyle, Charles, and Gay-Lussac are deductive consequences of axioms that state relations among moving "point-masses." Van der Waals made a natural modification of the kinetic theory. He replaced the concept "point-mass" with the concept "molecule of finite size." This modification enabled him to derive a formula—$(P + a\ /\ V^2)\ (V\text{-}b) = n\ R\ T$—that reproduces the pressure-volume-temperature behavior of a gas over a wider range of values than does the ideal gas law, [$P\ V = n\ R\ T$]. In the Van der Waals formula, *a* is a measure of intermolecular attractive forces and *b* is a correction term to account for the finite volume of the molecules.

Bohr's Correspondence Principle

If successful reduction is a prominent feature of the temporal development of science, then progress in science is much like the creation of an expanding nest of Chinese boxes. Niels Bohr championed this view of scientific progress. He maintained that the creation of ever-more-inclusive boxes conforms to the methodological requirements of the "correspondence principle."

The correspondence principle is an axiom of Bohr's early quantum theory of the hydrogen atom (1913). It stipulates that the hydrogen electron obeys the laws of classical electrodynamics in the limit as the radius of its orbit approaches infinity. The correspondence principle ensures that the equations of motion for discrete electron orbits pass over into the classical equations for the case of an electron no longer bound to the nucleus of the atom.[5]

Bohr insisted that the correspondence principle is more than an axiom of a specific theory. It is, in addition, a methodological directive. As such, it directs the theorist to build upon present theories in such a way that the successor theory and its predecessor are in agreement for that domain in which the earlier theory had been successful. The correspondence principle requires of each candidate to succeed theory T that

1. the new theory have greater testable content than T, and
2. the new theory be in asymptotic agreement with T in the region in which T is well confirmed.

Both Schrodinger's quantum theory and Einstein's special relativity theory satisfy the correspondence principle. In the limiting case in which the quantum of action may be neglected, Schrodinger's formulas for the probability distribution of the motions of quantum-mechanical particles reduce to the classical equations of motion. And in the limiting case in which the velocity of a system is negligible with respect to the velocity of light, Einstein's formulas also reduce to the classical equations of motion.

Ernest Hutten pointed out that in neither case does the more inclusive theory explicitly contain classical mechanics. Moreover, the correspondence principle may be satisfied despite the fact that a given term has a different meaning in each theory. For example, special relativity theory satisfies the correspondence principle even though "mass" is subject to rules of application that differ from those of classical mechanics. Relativistic "mass" is velocity-dependent. To satisfy the correspondence principle it suffices that, under specified limiting conditions, calculations made from the equations of special relativity theory approach asymptotically calculations made from the equations of classical mechanics.[6] The correspondence principle is a less restrictive criterion of successful theory-replacement than is reduction.

Karl Popper on Justification and Discovery

The Problem of Demarcation

The orthodox position sketched above was not without challenge in the early post-war period. Indeed, Karl Popper had presented an important alternative position even before the orthodox position had been developed. Popper outlined an original and distinctive philosophy of science in 1934.[1]

In so doing, he opposed the logical reconstructionist view of the philosophy of science. The logical reconstructionist withdraws hypotheses and theories from the flow of scientific activity, reformulates them in the categories of symbolic logic, and develops a logic of justification to evaluate them. Popper's position, by contrast, was that the contexts of discovery and justification are inextricably interrelated. The logical reconstructionist is wrong to attempt to freeze the action. Decisions about the acceptability of theories affect future theory-construction. And methodological decisions made in the search for more adequate theories in turn affect the way in which theories are evaluated. By reducing philosophy of science to an analysis of snapshots, the logical reconstructionist has missed the dynamic nature of science. According to Popper, it is precisely this dynamism, this susceptibility to revision, that distinguishes scientific interpretations from non-scientific interpretations.[2]

Popper called attention to the problem of specifying criteria of demarcation for scientific interpretations. Is a particular interpretation scientific? According to Popper, it depends on how it is defended. He noted that it is possible to defend a pet theory come what may. Various strategies are available to restore agreement between a theory and evidence that appears to count against it. The simplest strategy is to reject the evidence. But even if the evidence is accepted, the theory may be qualified or modified to accommodate the evidence.

For instance, consider the case of a subject who has exhibited "extra-sensory perception" (ESP) in card-predicting experiments on Monday and Tuesday but who fails to achieve predictive success on Wednesday. Does the evidence from Wednesday falsify the hypothesis that the subject possesses extra-sensory perception? Not necessarily. One can save the ESP hypothesis by "explaining away" Wednesday's results. Perhaps the subject had a headache. Or perhaps he sensed antagonism on the part of the person who turned over the cards. Or perhaps Wednesday's predictions were affected by a "forward displacement" such that, unknown to the subject, these predictions refer, not to the next card, but to the next-plus-one, or the next-plus-two, etc.

Suppose someone offers as a "scientific hypothesis" the claim that Jones has an extra-sensory predictive faculty, and then seeks to evade negative evidence by means of a range of strategies of the above type. At some point one becomes suspicious. The confrontation of hypothesis and evidence seems not to be genuine. Is there any evidence that could be specified that would be acknowledged to be decisive against the hypothesis?

Disputes may arise over strategies adopted to defend a threatened hypothesis. What one scientist considers to be a refinement, or natural extension, of a hypothesis, a second scientist may consider to be a misguided attempt to salvage a discredited hypothesis. An interesting case is the interpretation of the β-decay of radioactive nuclei. Measurements revealed that the energy of the nucleus before β-emission is greater than the sum of the energies of the daughter nucleus and the emitted electron. Scientists were loath to conclude that energy is not conserved in such processes. Wolfgang Pauli suggested in 1930 that a yet-undiscovered particle—the neutrino—was emitted along with the electron. The neutrino was presumed to carry off just that amount of energy required to insure conservation. It was not until 1953 that there was direct experimental evidence of the existence of this particle. Was the neutrino hypothesis in 1930 any less *ad hoc* than the hypothesis of "forward displacement of predictive power"?

Popper's position is that it is only with respect to the way in which hypotheses are defended that a satisfactory answer can be given to such a question. He recommended falsifiability as a criterion of demarcation for scientific interpretations. A hypothesis is scientific if, and only if, it is both logically possible and physically possible to falsify it. Present technical inability to design suitable tests need not disqualify a hypothesis.

For a hypothesis to qualify as scientific there must be some prospect for a confrontation with observational evidence. Nevertheless, the proponent of a hypothesis may respond to negative test results by making changes elsewhere in his assumptions. And scientists may judge him to be justified in making this move. Popper conceded that one cannot specify *a priori* the point at which the defense of a hypothesis ceases to be scientific. One can say, in advance, only that a person practices empirical methodology insofar as he exposes the entire interpretation to the possibility of being falsified, eventually.

Popper viewed the evaluation of scientific hypotheses and theories as a two-stage procedure. The first stage is the demarcation of scientific interpretations from non-scientific interpretations. The second stage is the evaluation of the acceptability of those interpretations that qualify as scientific.

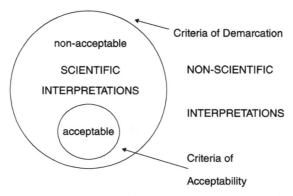

Figure 22 Popper on Demarcation.

Popper emphasized that demarcation is the logically prior problem. Before an interpretation is judged acceptable, it must qualify as scientific.

Popper did not originate this emphasis. It was implicit in Galileo's criticism of Aristotle's physics. Galileo realized that he could not disprove the Aristotelian thesis that "unsupported bodies move toward the earth in order to reach their 'natural place." But by restricting "scientific" interpretations to statements about "primary qualities" such as "position," "velocity," and "acceleration," Galileo excluded Aristotelian interpretations of the above type from physics.

However, it was Popper who promoted the problem of demarcation to center stage. He realized, as had Galileo, that there are certain interpretations that cannot be proved false, but which it is important to exclude from science. For instance, in the customary formulation of special relativity theory, the following claim may be excluded upon application of the falsifiability criterion: "The spectra of nebulae receding from the earth with a velocity greater than that of light have intense lines of ionized oxygen." Popper's emphasis on demarcation, like that of Galileo, coincided with a conceptual revolution in physics.

Demarcation is the first stage in the evaluation process. However, application of the falsifiability criterion establishes a demarcation only with respect to the method by which an interpretation is defended, Thus the first stage in the evaluation process (context of justification) involves the historical dimension of science (context of discovery). Popper used falsifiability in two ways:

1. as a criterion of demarcation in the context of justification; and
2. as a methodological rule in the context of discovery.

As a methodological rule, the falsifiability principle states that subsidiary rules selected to guide scientific discovery

> must be designed in such a way that they do not protect any statement in science against falsification.[3]

The Acceptability of Scientific Theories

Hypotheses that satisfy the falsifiability criterion fall within a range of permissible scientific discourse. It remains to assess their acceptability. Popper's view is that acceptable hypotheses have proved their mettle by withstanding tests designed to discredit them. Tests of hypotheses are like the destruction-tests that engineers perform to establish the strengths of materials.

Hypotheses are tested according to the following deductive pattern:

Hypothesis
Statements about antecedent conditions
_____

$\therefore$ Effect

If what is observed is the negation of the predicted effect, then the conjunction of premises is false.

Because it is only the conjunction of premises that can be falsified, a negative test result does not directly falsify a hypothesis. It may be the case that the antecedent conditions were not correctly described. For example, consider a test schema for the hypothesis that all blue litmus paper turns red in acid solution:

$(x) [(Bx \ \& \ Ax) \supset Rx]$
$Ba \ \& \ Aa$
_____

$\therefore Ra$

where $Bx = x$ is a case in which blue litmus paper is placed in solution

    $Ax = x$ is a case in which the solution is acidic, and

    $Rx = x$ is a case in which the paper turns red.

If what is observed is $\sim Ra$, then either the hypothesis or the statement of antecedent conditions (or both) is false. If there is reason to believe that there is no blue litmus dye on the paper, or that the solution is not acidic, then the hypothesis may be retained in spite of the negative test result. But the burden of proof is on the scientist who decided to retain the hypothesis. He must continue to expose the hypothesis to tests. Otherwise he is guilty of practicing non-empirical methodology.

Popper held that a test involves a confrontation between a hypothesis and a "basic statement." Basic statements are singular statements that specify the occurrence of an intersubjectively observable event in a certain region of space and time. To test a hypothesis one must accept as true a basic statement describing a test result. But basic statements themselves are not incorrigible. That a particular event occurred in a specific spatio-temporal region can be subjected to further tests. However, the testing of basic statements by reference to further statements must terminate at some point if the hypothesis itself is to be tested.

Popper acknowledged that, in the deductive testing of hypotheses, we *decide* to accept certain basic statements as true for the purpose of testing. There is an element of conventionalism in the deductive method of testing. Popper declared that

> the empirical basis of objective science has thus nothing "absolute" about it. Science does not rest upon rock-bottom. The bold structure of its theories arises, as it were, above a swamp. It is like a building erected on piles. The piles are driven down from above into the swamp, but not down to any natural or "given" base; and when we cease our attempts to drive our piles into a deeper layer, it is not because we have reached firm ground. We simply stop when we are satisfied that they are firm enough to carry the structure, at least for the time being.[4]

Popper maintained that the acceptability of a hypothesis is measured by the number and the severity of tests it has passed.[5] The severity of a test is difficult to measure. It depends on such factors as the ingenuity of the experimental arrangement, the accuracy and precision of the results achieved, and the relationship of the test result to rival hypotheses. In addition, much depends on how tightly the hypothesis tested is bound within a theoretical framework. If the hypothesis is so tightly bound that a negative test result could be accommodated only by a drastic revision of the system, then the test is a severe one. By this standard, the discovery that light rays from the stars are "bent" by the sun, as predicted by Einstein's theory of general relativity, is a test of high-level severity. By contrast, the discovery of yet another raven whose feathers are black is a test of low-level severity.

The Corroboration of Hypotheses

Popper's catalog of difficulties that beset the evaluation of hypotheses is extensive. Nevertheless, he made two suggestions about a quantitative measure of the acceptability of hypotheses.

His first suggestion was that the concept of "logical probability" is not a suitable measure of acceptability. In the first place, to take the probability of a hypothesis to be the ratio "favorable cases/total cases" within a class of evidence statements would be to assign a probability of 1/2 to a generalization refuted by every second statement in the evidence class. Of course, this would be a crude way to assign probabilities. But Popper urged that the very idea of measuring acceptability by reference to logical probability is misguided. Proponents of inductive logic often focus on the ratio

$$\frac{\text{number of statements of favorable instances}}{\text{total number of statements of relevant instances}}$$

According to Popper, what counts are *tests* and not mere instances. To take a simple example, discovery of a black raven living in the polar region would be a more important favorable instance of the raven hypothesis than discovery of yet another temperate-region black raven. Inductive logic is not well suited to measure the severity of tests.

Moreover, given a set of statements that record positive evidence *e*, the "most probable" hypothesis is that hypothesis that implies just *e* and nothing further. "Highly probable" hypotheses, in the sense of "logical probability," are those hypotheses that venture least. Since an important aim of science is to maximize the content of hypotheses, "logical probability" is not a suitable measure of the acceptability of hypotheses.

Popper's second suggestion was to introduce the phrase "degree of corroboration" as a measure of how well a scientific theory has withstood tests. He suggested that "degree of corroboration" is an index of the fitness for survival of scientific theories. Its magnitude increases with the increasing severity and diversity of tests successfully passed. A highly acceptable theory is one that has passed tests in widely different fields of application. For instance, Einstein's theory of general relativity has passed tests that include the shift of spectral lines in vibrating atoms, the bending of light rays that pass close to the sun, and the resolution of a discrepancy in the observed orbit of Mercury.

Popper pointed out that his position on corroboration reflects the "growth by incorporation" that often characterizes scientific progress. In such cases one theory is superseded by a second theory which contains the first theory, or an approximate version of it, and which has additional testable consequences.[6] This is the relationship between Newton's mechanics and Einstein's special relativity theory, for example. Popper emphasized that a principal aim of science is the formulation of theories of increasing scope.

On the logical reconstructionist view of science the acceptability of a hypothesis is determined by its logical relationship to statements recording evidence. If this is the case, then it does not matter whether the hypothesis or the evidence is presented first. Hempel expressed the situation as follows:

> from a logical point of view, the strength of the support that a hypothesis receives from a given body of data should depend only on what the hypothesis asserts and what the data are: the question of whether the hypothesis or the data were presented first, being a purely historical matter, should not count as affecting the confirmation of the hypothesis.[7]

Popper insisted that the temporal relationship between hypothesis and evidence cannot be disregarded. In particular, the acceptability of a hypothesis is enhanced if it proves applicable to a range of phenomena not initially taken into account. Popper supported John Herschel's position that a display of "undesigned scope" provides important support for a hypothesis.[8] For instance, Einstein's extension of Planck's quantum hypothesis to the photoelectric effect gave impressive support to the hypothesis of energy quantization. It was judged important that this extension revealed that Planck's hypothesis had an "undesigned scope." Had Planck initially applied the hypothesis of the quantization of energy to both black-body radiation and the photoelectric effect, the hypothesis presumably would have received less support.

Popper maintained that the acceptability of a theory depends on its temporal relations to confirming instances.[9] Emphasis on the importance of undesigned scope is an aspect of Popper's insistence that the contexts of justification and discovery are interdependent.

Orthodoxy Under Attack

Doubts About the Two-Tier View of Scientific Language

Feigl's Attack on the Two-Tier View

Herbert Feigl was an early defector from the ranks of orthodoxy. In a paper published in 1950, Feigl questioned the adequacy of the orthodox position.[1] Probing for weaknesses in the orthodox position, he analyzed the distinction between observation and theory, and stressed the role of existential hypotheses in the growth of scientific knowledge.

The Observation–Theory Distinction

Feigl sought to blur the distinction between observation reports and theoretical claims. He did so in order to discredit the orthodox position that there exists a theory-independent observation language to which one may appeal to decide between competing theories. He pointed out, in the first place, that theories are not accepted or rejected upon straightforward appeal to some inviolable set of observation reports. That one uses a theory to predict O, and then observes $\sim O$, need not lead to abandonment of the theory. On the contrary, divergences between observation reports and predictions drawn from theory often are "corrected from above" by the introduction of further theoretical considerations. One of Feigl's illustrations is the use of the theory of perturbations developed from Newton's theory of mechanics to restore agreement between Kepler's laws and the observed motions of the planets. By introducing the perturbing gravitational forces of Jupiter and Saturn, for instance, the failure of Mars to conform to a Kepler's-law elliptical orbit can be explained. We do not discard Kepler's laws on account of this "correction from above."

Feigl also maintained that the "confirmation basis" of a theory can be specified only upon appeal to general theoretical principles. He called attention to the interdependence of theory and observation in the determination of the presence of a field in a particular region. That a magnetic field is present in a certain region may count as confirming evidence for electromagnetic theory. And the deflection of a magnetometer needle presumably is evidence for the existence of a magnetic field. However, this evidence cannot be characterized independently of theoretical considerations. Deflection of the needle counts as evidence only if the instrument has been constructed and calibrated in a certain way. Questions about proper construction and calibration involve an appeal to electromagnetic theory itself.

Feigl did not make clear whether he wished to claim

1. that the observation statements that confirm a theory cannot be formulated without reference to theoretical statements of that same theory, or only
2. that the observation statements that confirm a theory cannot be formulated without reference to some general theoretical considerations.

One might concede that weaker claim (2) is correct, and yet insist that observation statements can provide independent support for a particular theory. However, Feigl's magnetic field example can be used to support a stronger claim (1). Indeed, Feigl did assert that

> what epistemologically must be looked at as the confirmation basis of the hypothetical construction, will in the full-fledged theory be given a place within the cosmos of which the theory treats.[2]

If the stronger claim is correct, then no theory-neutral observation language is available and the two-tier view of theories must be discarded.

The Status of Existential Hypotheses

Feigl criticized supporters of the orthodox view for failing to emphasize the role of existential hypotheses in the growth of science. Hypotheses that assert the existence of neutrinos, viruses, and magnetic fields have been of great importance in recent science. Feigl claimed that on the orthodox view, the meaning of an existential hypothesis is given by the set of observation statements that can be deduced from premises that include the hypothesis. However, this view provides insufficient rationale for the numerous experimental investigations undertaken to establish the existence of such entities and properties. Feigl concluded that an existential hypothesis has a "surplus meaning" over and above that which is established by its confirmation relations. He suggested that a "semantical realism" can account for this "surplus meaning" of existential hypotheses, presumably in terms of some additional "factual reference" of these hypotheses.

However, Feigl was not clear what this "factual reference" is. He was concerned to avoid a "metaphysical realism" that populates the universe with weird entities. "Semantical realism," he wrote,

> is free from the dangers of metaphysics precisely because it does *not* prescribe anything at all about the *nature* of the designata of our theoretical constructs. It is concerned only with the most abstract and formal features of the semiotic situation. There is no danger that the wish for picturization, so strong in the older metaphysical form of realism, will dictate the application of the categories of common sense to domains where they are notoriously out of place.[3]

It would seem that Feigl's "semantical realism" requires of an hypothesis, such as "electrons exist," that

1. the hypothesis is not reducible without loss of meaning to statements about Geiger counter clicks, tracks in cloud chambers, and the like, and
2. the "surplus meaning" of the existential hypothesis does not include the claim that there exist "tiny billiard balls" analogous to medium-sized spheres that can be apprehended directly by the senses.

But then what is this "surplus meaning"? Why did Feigl refer to his position as "semantical realism"?

Replies to Feigl's Attack

In the same issue of *Philosophy of Science*, Hempel and Nagel replied to the challenge of semantical realism.[4]

Hempel addressed the question of the surplus meaning of existential hypotheses. He conceded to Feigl that the meaning of an existential hypothesis is not equivalent to the set of observation statements that report the outcome of possible tests of the hypothesis. This is because

> the meaning of a statement in the language of science is reflected in the totality of its logical relationships (those of entailment as well as those of confirmation) to all other sentences of that language.[5]

Thus the meaning of "positrons exist" is not exhausted by the relations of confirmation by which the hypothesis is linked to observation statements. The hypothesis also is embedded in a network of theoretical assumptions.

Hempel agreed with Feigl that existential hypotheses have a "surplus meaning." However, he denied that this surplus meaning is some "factual reference" of the constructs involved. It is, rather, the totality of logical relationships into which the hypothesis enters over and above the confirmation relations in which the hypothesis is linked to observation statements.

In his reply, Nagel emphasized the instrumental role of hypothetical constructs. What is important, he urged, is the way in which statements about hypothetical constructs establish connections between classes of observation statements, and not any symbolic representation of some supposed realm of objects.

According to Nagel, the "surplus meaning" of an existential hypothesis derives from the possibility that there is more evidence for it than is currently available. This possibility of additional confirming evidence may depend on logical relations between the hypothesis and other theoretical assumptions.

C. W. Churchman challenged Feigl to show how it makes a difference whether one adopts semantical realism or the orthodox position.[6] Churchman questioned the "cash value" of the semantical realist interpretation of theories. Does it have consequences for the conduct of scientific inquiry, or is it merely a colorful way of speaking about theories?

In a "Reply to Critics,"[7] Feigl took up the challenge set by Churchman. He insisted that semantical realism and the orthodox position are rival interpretations of theory-construction, and that it does make a difference which is adopted. This is made clear by a consideration of the role of existential hypotheses in science. In the kinetic theory of gases, for example, assertions are made about the velocities of individual molecules. A hypothesis of the theory is that molecules exist.

Many years after the theory had been formulated, Born and Stern devised experiments to measure the velocities of individual gas molecules. This development is easily explained on the semantical realist interpretation that assigns a "factual reference" to the existential hypothesis. The investigators presumably believed that there are such things as molecules and that an attempt to measure their properties is a significant enterprise. Feigl maintained that the situation is different on the orthodox position. He noted that

> since "constructs" like the mass and the velocity of individual molecules are expressly viewed (by the phenomenalists) as "nothing but" parameters in an abstract model, he could not *on this interpretation* have predicted with any appreciable probability the outcome of such experiments as that of Born and Stern.[8]

Feigl weakened his case, in the above passage and elsewhere, by arguing against "phenomenalism" as if Hempel and Nagel defended such a position.

However, neither Hempel nor Nagel subscribed to a phenomenalism that holds that the meaning of a theoretical claim is the totality of observation reports that would confirm it. For instance, neither Hempel nor Nagel held that assertions about the postulated micro-structure of a gas are equivalent to assertions about its observed (or observable) macroscopic behavior.

Nagel, in particular, took pains to distinguish the two following questions:

1. Can theoretical statements be translated into the language of direct observation?
2. What is the instrumental role of theoretical statements in the conduct of inquiry?

He maintained that if the "translation" of a theoretical statement T into an observational language requires that T is logically equivalent to some finite set of statements O_i that report the results of observations, then the translatability requirement is not fulfilled. Nagel pointed out that insofar as this translatability thesis is central to phenomenalism, neither he nor Hempel were phenomenalists, and Feigl's criticism of this position was not to the point.

Nagel maintained that it is only the second of the above questions that is in dispute. He noted that an "operationalist" theory of science successfully explains how statements about hypothetical constructs establish connections between classes of observation statements. For instance, statements about the motions of individual gas molecules are used in premises of deductive arguments, the conclusions of which describe the macroscopic behavior of gases.

Although Feigl hardly could be said to have won a victory for semantical realism, he did locate a weakness in the orthodox position. Proponents of the orthodox position had been preoccupied with universal generalizations, their confirmation, and their role in explanations. Existential hypotheses had not been much discussed. Feigl called attention to this oversight. He challenged supporters of the orthodox position to account for the importance of these hypotheses in the growth of scientific knowledge.

Hesse's Analysis of the Testing of Theories

Mary Hesse delivered another challenge to the orthodox view in 1958.[9] Hesse asked whether the two-tier view was adequate to account for the testing of scientific theories. She declared that if an observation report is to be a test of a theory, then the meaning of that report cannot be independent of that theory. Consider a test of the wave theory of light. One of the derived laws of the theory is Snel's Law—sin i / sin r = constant, where *i* and *r* are the angle of incidence and the angle of refraction of a beam of light. Hesse pointed out that if *i* and *r* were uninterpreted symbols, we would not know how to test the law. Perhaps *i* and *r* stand for the angle between earth, Mars, and Venus on two occasions. It is only the *interpreted* formalism that can be tested. We have to know how to link the symbol, *via* rules of correspondence, to operationally defined magnitudes before we can perform a *bona fide* test of the theory.

Suppose we observe that a magnifying glass, suitably placed between the sun and a piece of paper, ignites the paper. Does this count as a test of the theory of refraction? According to Hesse, it does only if we give a theoretical interpretation to the event. If we simply describe the geometry of the arrangement, the color of the flame, and the charred remains, there would be no reason to connect this description with the theory of refraction. But if we extend Snel's Law to cover refraction at curved surfaces, and refer to the electromagnetic theory of light, then we can interpret the concentration of light rays at a point on the paper as a process of energy transfer that raises the temperature of the paper above its kindling temperature. Thus interpreted, we can take this as confirming evidence for the theory of refraction.

Hesse concluded, from considerations such as these, that the two-tier view is incorrect. It is not theory-independent observation statements that confirm scientific theories. Rather, theories are tested by observation statements that are given a theoretical interpretation.

Alexander's Defense of the Two-Tier View

Hesse's paper drew a response from Peter Alexander.[10] Alexander sought to defend the two-tier view by shifting emphasis from the testing of theories to the antecedent question of how a theory accounts for a type of phenomenon. Confronted with a puzzling phenomenon, a scientist may construct a theory to explain it. For an explanation to be acceptable, the description of the phenomenon must not be given in terms of the concepts of the theory constructed to explain it. For example, if one "describes" what takes place in a cloud chamber as the production of an electron-positron pair, then one already has loaded the description in such a way that it would be superfluous to "explain" this phenomenon by citing the transformation of energy into an electron and a positron.

According to Alexander, the point at which a theory-neutral observation language is required is the initial description of the phenomenon to be explained. He emphasized that what is important is that the initial description not make reference to the theory used to explain it. It is only with respect to the explanatory theory in question that the description must be theory-neutral. The description may be given in the concepts of other theories. For example, it is unobjectionable to describe atomic spectra in concepts drawn from the wave theory of light ("wavelength," "frequency," et al.), and then to explain the spectra by hypothesizing atomic transitions within atoms.

Alexander maintained that Hesse had failed to distinguish "understanding an observation statement" from "understanding that an observation statement is a test statement for a particular theory." The former does not require knowledge of the theory in question, even though the latter does. A description of the burning of a piece of paper beneath a magnifying glass can be understood without reference to electromagnetic theory. That this description is a test statement for electromagnetic theory depends on the meaning of the theory.

Alexander claimed further that a description can be a test statement for a theory only if it can be understood without reference to that theory. This claim is a cornerstone of the orthodox view. He argued that

> there must, of course, be connections of meaning between phenomenal statements and the theory that explains them, otherwise the theory would not explain them, but this does not imply that the meaning of a statement is altered when a theory is found to explain it or that it cannot be understood without knowledge of that theory.[11]

Robert Boyle's understanding of "P V = k" was as adequate as that of James Clerk Maxwell, even though Maxwell could give a better explanation of why this relationship holds.[12]

Achinstein on the Limitations of the Two-Tier View

Peter Achinstein called attention to the ways in which the observable–non-observable distinction is drawn in practice. Suppose a biologist examines a tissue sample successively with the naked eye, under a microscopic, under a microscope after staining and fixing, and under an electron microscope. In which of these cases can he be said to have "observed" the tissue?[13] Achinstein's answer was that it all depends on what contrast is intended between "observable" and "non-observable."[14]

One might maintain that the tissue is "observed" only by unaided eyesight. What is seen through the lens of a microscope is not the tissue itself but an optical image. Alternatively, one might accept microscopic observations but exclude the viewing of fluorescent-screen images produced by electron-beam bombardment in an electron microscope. Achinstein emphasized that there are numerous ways to subdivide terms into "observables" and "non-observables." A classification adequate for one purpose may be inadequate for another purpose.

Achinstein noted that there is an alternative approach to the classification problem. One may seek to isolate those terms that are indisputably theoretical. Achinstein examined proposals that had been put forward by N. R. Hanson and Gilbert Ryle.

Hanson suggested that certain scientific terms are "theory-laden." Such terms "carry a conceptual pattern with them."[15] For instance, to speak of certain dark spots on the surface of the moon as "craters" is to use a theory-laden term. To use this term is to impress an organizing pattern on what is observed. By contrast, to speak of the dark spots as "concavities" is to adopt a more neutral, or "phenomenal," standpoint.

Achinstein conceded that such contrasts can be drawn. He insisted, however, that these contrasts are context-dependent. For example, there are contexts in which use of the term "concavities" also imposes organization upon initially puzzling data. Achinstein noted that Hanson had admitted as much. Hanson wrote that

> it is not certain that certain words are absolutely theory-loaded whilst others are absolutely sense-datum words. Which are the data-words and which are the theory-words is a contextual question.[16]

Achinstein insisted that if a given word can be theory-laden in one context and a phenomenal term in another context, then one cannot appeal to "theory-ladenness" to demarcate theoretical terms.

Nevertheless, perhaps one could rope off theoretical terms by stipulating that a term is theoretical if, and only if, its meaning is given by the principles of some theory. Ryle had suggested that a criterion of this sort might be used to make judgments about comparative theory-dependence. He drew a contrast between the term "straight flush" and the term "queen of hearts." The former is theory-dependent (poker-dependent) in a way in which the latter is not.[17] Similarly, "intelligence quotient" is theory-dependent in a way in which "test score" is not.

Achinstein granted that such comparisons can be made. He emphasized, however, that Ryle's approach does not establish a demarcation for the theoretical terms of the sciences. Achinstein's presentation suggests that he is criticizing a proposal to demarcate theoretical terms. However, Ryle at no point claimed that the vocabulary of the sciences could be subdivided neatly into theoretical terms and non-theoretical terms. Ryle merely sought to compare the different use of terms within specific theories.

Nevertheless, Achinstein advanced several reasons why Ryle's criterion could not be used to support the two-tier view of the language of science. One reason a subdivision cannot be achieved is that there are numerous *prima facie* non-theoretical terms, the meanings of which are given by "the rules of the game." For example, terms such as "on," "off," and "dial" are "instrument-dependent" in much the same way that "straight flush" is "poker-dependent." One has to have some knowledge of instruments to understand the meanings of these terms. However, these "instrument-dependent" terms are not the sort of terms that orthodox theorists classify as theoretical terms.[18]

An additional difficulty is that certain terms occur in a number of theories. "Temperature" is a case in point. We speak of the "temperature at the center of a star" as well as "the temperature of the water in a glass." The "temperature at the center of a star" is a highly theoretical term. Its meaning depends on hypotheses about nuclear energy generation, the variation of chemical composition with radius, the variation of density with radius, and other factors. In the spirit of Ryle's discussion of comparative theory-ladenness, the "temperature of the water in a glass" is not theory-dependent to the same degree. To speak correctly about "temperature" one would need to specify the theory in which it is embedded.

Achinstein observed, moreover, that a term may be judged to be theory-dependent (in Ryle's sense) in one theory and theory-independent in another theory. For example, "mass" may be judged to be theory-dependent in Newtonian mechanics, but theory-independent in Bohr's theory of the hydrogen atom. Which terms appear in a list of theoretical terms depends on which theories are considered. Achinstein concluded:

> I have considered the widespread doctrine that there exists a fundamental distinction between two sorts of term employed by scientists. On one view the distinction rests on observation; on another theory-dependence ... [Neither approach] will generate the very broad sort of distinction so widely assumed in the philosophy of science.[19]

Feyerabend's Claim that the Observation Language Is Theory-Dependent

It was the orthodox position that the theoretical level of scientific language is parasitic on the observational level. Observation reports would be unchanged if there were no theories, but theories could not exist apart from observation reports. Paul Feyerabend suggested that the relationship is otherwise. It is observation reports that depend for their existence on theories. He declared that

the interpretation of an observation-language is determined by the theories which we use to explain what we observe, and it changes as soon as those theories change.[20]

Feyerabend created the following example to illustrate this point.[21]

A group of scientists ascribe color-predicates—$P_1, P_2, P_3, \ldots$—to self-luminescent objects with names $a, b, c, \ldots$ When questioned about the meaning of a claim such as "a is P_1," the scientists agree that P_1 is a property of the object named, and that the object has P_1 independently of its relation to any observer. Now suppose that a dissident scientist claims that the color recorded by an observer depends on the relative velocity of object and observer. Observation report "a has P_1" has been reinterpreted. According to the dissident scientist, to ascribe a color-predicate to an object is to state a relation between object and observer, and it no longer is meaningful to speak of the color of an unobserved object.

If Feyerabend is correct about the theory-dependence of observation reports, then several important consequences follow. One consequence is that the distinction between theoretical statements and observational statements is a pragmatic distinction, and not a distinction based on a difference in logical type. A second consequence is that, since all terms are theory-dependent, there is no special "problem of theoretical terms." And a third consequence is that there can be no "crucial experiments" in science. An observation report is a statement that has received an interpretation from some theory. To accept an observation report as deciding the issue between competing theories is to accept the truth of the theory that provides its interpretation. The observation report is subject to interpretation at the hands of some new theory, and the reinterpretation may change its status as a test of other theories. Hence, an observation report can be no more than "crucial provided that one accepts the particular theory that interprets it."

Quine's Criticism of Empiricist Assumptions

A still more incisive challenge to the two-tier view of scientific language was issued in 1950 by Harvard philosopher Willard van Orman Quine. Quine suggested that a scientific theory is rather like a field of force that is subject to constraints provided by experience.[22] Echoing a thesis that had been defended by Pierre Duhem,[23] Quine declared that a recalcitrant experience can ... be accommodated by any of various alternative reevaluations in various alternative quarters of the total system.[24]

Given a conflict between theory and experience, we may, and usually do, choose to adjust those parts of the force field at the periphery. By so doing, we reestablish agreement with observations by making changes that have minimum repercussions within the theory. But we are not forced to respond in such a conservative manner. Instead we may make changes deep within the force field, changes that greatly affect all regions of the field. An example of this more drastic accommodation is Hans Reichenbach's proposal to recast quantum theory in the categories of three-value logic.[25] Such a change would produce extensive repercussions within the theory. Regardless of the strategy selected to restore agreement with observations, any given statement within the theory can be retained as true provided that sufficiently drastic adjustments are made elsewhere in the system.[26]

One important aspect of the "Duhem-Quine thesis" is that there is no inflexible division of the statements of a theory into those that are analytic (truth-status independent of empirical considerations) and those that are synthetic (truth-status contingent upon empirical considerations). This poses a threat to the orthodox position on the structure of scientific theories. On Hempel's safety-net view, for instance, undeniably synthetic observation statements provide firm support for the poles (rules of correspondence) that support the net (axiom system). But if Feyerabend and Quine are correct, there are no theory-independent observation reports. Rather, the observation reports that support a theory are set within a theoretical context.

Quine's essay provoked a number of responses. One group of respondents challenged Quine's position on the analytic–synthetic distinction. A second group of respondents challenged Quine's position on the nonfalsifiability of individual hypotheses.

The Debate over the Analytic–Synthetic Distinction

Herbert Feigl insisted, against Quine, that the analytic–synthetic distinction (ASD), is indispensable for intelligible communication. He maintained that intelligible communication presupposes acceptance of rules of inference, acceptance of rules of inference presupposes acceptance of a language in which words have contextually fixed meanings, acceptance of such a language presupposes the existence of synonymies and logical equivalences within the language, and the existence of synonymies presupposes the existence of statements that are analytic within the language.[27]

Feigl pointed out, in addition, that the ASD is the basis of the distinction between an axiom system and its application to experience. For example, the distinction between pure geometry and physical geometry is a distinction between a set of statements that are analytically true and a set of statements that are synthetically true. The ASD also is implicated in the distinction between deductive validity and inductive support.

None of this is decisive against Quine. Quine had not denied that, in practice, we do protect some statements from possible revision by making changes elsewhere in the linguistic system. Feigl's complaint was just that—a complaint. However, it was given support by the arguments of H. P. Grice and P. W. Strawson.

Grice and Strawson professed to be uncertain as to just what Quine wished to claim. They suggested that if the claim is that no analytic–synthetic distinction exists, then Quine has not proved his case. Grice and Strawson observed that in ordinary usage, statements like "one yard is 36 inches" and "every bachelor is unmarried" are classified as analytic. Statements like "Mount Everest is higher than Mount Hood" and "every bachelor is wealthy" are classified as synthetic. Grice and Strawson emphasized that

> if a pair of contrasting expressions are habitually and generally used in application to the same cases, where these cases do not form a closed list, this is a sufficient condition for saying that there are *kinds* of cases to which the expressions apply; and nothing more is needed for them to mark a distinction.[28]

Perhaps Quine's point is only that there is an ASD but that it has been misunderstood by philosophers. If this is the point at issue, then what would constitute a correct understanding of this distinction? Quine demanded a "clarification" of "analyticity" that breaks out of the network below:

"analytic"_____"synonymous"

/   \

"self-contradictory"_____"necessary"

According to Grice and Strawson, this is an unreasonable demand. They pointed out that many important concepts are embedded in a network of relations to other concepts. An example from the law is:

"contract"_____"property"

/ |

"person"_____"breach of contract"

Quine appeared to argue that because no satisfactory "clarification" has been given of "analyticity," the concept is not meaningful. However, Quine's standard of "clarification" is too severe. He demanded a set of necessary and sufficient conditions for the application of "analytic" such that no reference is made to the other terms of the network.

Grice and Strawson insisted that one can teach the use of one of the terms in a network of terms without being able to specify a set of necessary and sufficient conditions for application of the term. For instance, by citing appropriate examples, one can teach the use of "analytic" or "logically impossible." The test of understanding is correct usage of the term in further contexts.

Grice and Strawson sought to remove the threat posed by Quine's attack on the ASD by distinguishing two questions:

1. Are there linguistic forms that express analytic propositions regardless of the conceptual scheme in which they are embedded?
2. Are there linguistic forms within some conceptual scheme that express analytic propositions?

Grice and Strawson agreed with Quine that the answer to (1) is "no." There is no statement that in principle is immune from revision. Nevertheless a distinction can be made between analytic and synthetic statements within a conceptual scheme. The ASD may reflect usage within a particular language. So long as language users accept the stipulation that "one yard equals 36 inches," for example, this stipulation is an analytic truth. The answer to (2) is "yes." Quine's arguments on behalf of a negative answer to (1) do not provide support for a negative answer to (2).

To accept an affirmative answer to (2) is to accept a distinction between two ways in which a statement may be rejected. The first way is by admitting its falsity. The second way is by recognizing that the meanings of concepts have changed. In cases of meaning-change, a statement at one time held to be true subsequently may be held to be false.

In articles published in 1961 and 1962, Grover Maxwell applied the distinction between "analytic in any context" and "analytic in a particular context" to scientific theories.[29] Maxwell agreed with Quine, and Grice and Strawson, that no statement is "analytic-in-itself." He maintained, however, that there are analytic statements within scientific theories. Analytic statements are statements that stipulate the meanings of certain of the terms of the theory. There are various ways in which meanings may be stipulated. Maxwell listed the following:

1. explicit definitions—e.g., "H = E + P V"; "v_{ave} = d / t";
2. partial explicit definitions—e.g., taxonomic definitions by genus and difference;
3. implicit definitions of theoretical terms by a subset of the axioms and rules of correspondence of a theory—e.g., v_i in the kinetic theory of gases; Ψ in quantum mechanics; and
4. other stipulations—e.g., "nothing is both red and green at the same time"; "the one-way velocity of light is a constant that is independent of direction"; "nothing behaves both as a particle and a wave in the same experimental arrangement."

According to Maxwell, to ask whether a statement within a theory is analytic is to ask about what has been stipulated within a theory. Maxwell emphasized that a particular theory may be formulated in diverse ways. Consider Newtonian mechanics. Following Mach, one may take the second axiom—"F = ma"—to be analytically true. In this formulation, Hooke's Law—"F = –k Δ x"—is a synthetic statement whose truth is contingent upon experience. Alternatively, one may take Hooke's Law to be analytically true, in which case "F = ma" is a contingent empirical generalization. In either formulation, the term "force" is implicitly defined by a subset of axioms and correspondence rules. But a different subset is involved in each formulation. Thus, a statement that is analytic in one formulation may be synthetic in another.

A decade after Quine's initial challenge to orthodoxy, Hilary Putnam appraised the course of the debate over the analytic–synthetic distinction.[30] He chided Quine's critics for "refutations" devoted to the citing of examples of "analytic" statements. Putnam did not name the critics who were guilty of this type of "refutation," but he did exempt from this derogatory appraisal the paper by Grice and Strawson.

Putnam's own evaluation of Quine's challenge was that Quine was wrong to deny the existence of the ASD, but right to emphasize that the ASD had been misused by philosophers. According to Putnam, there is a distinction to be made between analytic statements and synthetic statements. However, it is a rather unimportant distinction, because analytic statements are not of much interest. Putnam recommended a model in which statements reflect varying degrees of linguistic convention and systematic import (see Figure 23).

Analytic statements reflect a maximum of linguistic convention and a minimum of systematic import. They anchor the scale, as it were. Because they lack systematic import—they are not of much interest. At the other end of the

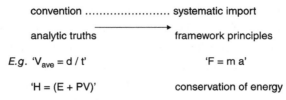

Figure 23 Putnam on Linguistic Convention and Systematic Import.

scale are "framework principles." Each framework principle is used in a variety of explanatory arguments.

Putnam emphasized that it is a mistake to classify as "analytic" both linguistic conventions and framework principles. Only linguistic conventions are properly termed "analytic."

Putnam brought out the difference between the two kinds of statements by pointing out the ways in which they may be overthrown. Analytic statements may be overthrown only by the

> trivial kind of revision which arises from unintended and unexplained historical changes in the use of language.[31]

That "all bachelors are unmarried" one day may held to be false, but only if "bachelor" or "marriage" undergoes a drastic change of meaning in the evolution of our language.

Framework principles, by contrast, are subject to refutation in a stronger sense. A framework principle cannot be overthrown by an isolated experimental result. But it can be overthrown by the success of a conceptual system that incorporates a principle that is incompatible with it. "$E = 1/2\ m\ v^2$" is a framework principle that has been overthrown. In Einstein's relativity theory, the corresponding expression for "energy" is "$E = m\ c^2 + 1/2\ m\ v^2 + 3/8\ m\ v^4 + \ldots$" The conservation of mass and the conservation of parity are other framework principles that have been overthrown.

Framework principles are assertions about "law-cluster" concepts. "Force" and "energy," for instance, participate in a number of laws. Because of this participation, a rejection of "$F = m\ a$" or "$E = 1/2\ m\ v^2$" produces repercussions throughout physics. Of course, scientists should be open to the possibility of revising framework principles. It would be a mistake to treat framework principles as if they were analytic truths.

The rejection of an analytic statement, by contrast, would ripple no theoretical waters. Consider the stock example "all bachelors are never-married adult male human beings." The only true universal conditional of the form "all bachelors are . . ." is the stock example itself (and its logical consequences). Since there are no other true statements of the form "all bachelors are . . .," the fact that the stock example is immune from revision cannot impede the growth of our knowledge.

But how can we be sure that there are no universal laws to establish "bachelor" as a "law-cluster" term? Putnam conceded that the existence of such laws is logically possible. He maintained, however, that the absence of any good reason to affirm the existence of such laws is itself a good reason to conclude that such laws do not exist.[32]

Grunbaum on the Duhem-Quine Thesis

Quine's principal challenge to orthodoxy was at the point of the falsifiability of an individual hypothesis. He acknowledged Pierre Duhem as the initial proponent of this thesis. Duhem had maintained that an observation report cannot falsify a hypothesis.

Consider the following relation:

$$(H \ \& \ A) \rightarrow O,$$

where H is the hypothesis,

A is a set of auxiliary hypotheses (statements of boundary conditions, statements of initial conditions, hypotheses about the workings of instruments, and other hypotheses tacitly assumed in the application of H in the given contest),

O is an observation report, and

$\rightarrow$ stands for the relation of logical implication.

Given that $\sim O$ is the case, what is falsified is the conjunction (H & A).

This is a point of logic, since no valid deductive argument can have true premises and a false conclusion. But since a conjunction is false if either of its conjuncts is false, one way that "(H & A)" could be false is that A is false and H is true. The truth of $\sim O$ does not establish the falsity of H itself.

Quine maintained, in addition, that regardless of the specific content of $\sim O$, it always is possible to find another set of auxiliary assumptions A^* such that

$$(H \ \& \ A^*) \rightarrow \sim O.$$

This strategy retains H as a premise in an explanatory argument, in this case with $\sim O$ as conclusion. By selecting an appropriate A^*, the scientist can retain H in the face of the *prima facie* falsifying evidence $\sim O$.

In articles published between 1960 and 1963, Adolf Grunbaum argued against the Duhem-Quine thesis.[33] As stated above, Quine's thesis is trivially true. Grunbaum observed that by selecting "A* = (H & ~O)," the desired implication is secured. If the Quine thesis is to be non-trivial, this kind of adjustment in the auxiliary hypotheses must be prohibited.

Another requirement for the non-triviality of the Quine thesis is that the language in which H, O, and A are formulated is "semantically stable." One would be defending the Quine thesis in a trivial manner if he secured the nonfalsifiability of H by an *ad hoc* reinterpretation of the meanings of concepts. Grunbaum's example is to shift the intension of "ordinary buttermilk" to that of "arsenic" in order to "save" the hypothesis that "ordinary buttermilk is toxic to humans."[34] The Quine thesis is fulfilled trivially if it is permissible to alter the meanings of concepts for the sole purpose of creating an A^* that, together with H, implies $\sim O$.

Grunbaum did not elaborate on the requirement of "semantic stability." Clearly concepts such as "mass," "temperature," and "electron" have undergone changes of meaning as one theory replaced another. We believe that such changes were justified. On the other hand, phlogiston theorists were not justified to alter the meaning of "chemical substance" so as to permit a "substance" to have a negative weight.

Grunbaum conceded that he could not give a set of sufficient conditions for the non-triviality of the Quine thesis. But he insisted that there are reasons to reject a non-trivial version of the thesis.

In its non-trivial form, the Quine thesis is that for each case of *prima facie* falsification "$[(H \& A) \rightarrow O \& \sim O]$," it is possible to find a non-trivial A^* such that

$$(\exists A^*)\,[(H \& A^*) \rightarrow \sim O]$$

Grunbaum emphasized that there is no logical necessity that there should be a non-trivial A^* available to satisfy the above condition. He maintained, in addition, that in at least one case, a hypothesis *is* falsifiable separately by adverse empirical evidence. He held that physical geometry affords a counter-case to the Quine thesis. Grunbaum directed his analysis to Einstein's version of this thesis.

Einstein had argued that any attempt to falsify a physical geometry must take into account the thermal, elastic, and electromagnetic deformations exhibited by whatever solid rods are used to make measurements. Because of the presence of these deforming forces experimental observations can falsify only the conjunction of the geometrical hypothesis and physical hypotheses about deforming forces., viz.,

$$([(G \;\&\; P) \to O] \;\&\; \sim O) \to \sim(G \;\&\; P)$$

where G is a geometrical hypothesis (e.g., the euclidean theorem that the sum of the angles of a triangle is 180°, for which semantic rules have been specified to link terms such as "line" and "angle" with measuring operations), and P is a physical hypothesis about deforming forces that may affect the measuring instruments.[35]

Einstein noted that the laws governing deforming forces themselves presuppose a geometry, since a determination of strains and stresses depends on the calculation of areas and volumes. Before rods corrected for deformations can be used to test the physical geometry in question, a physical geometry (usually euclidean geometry) already has been employed. For this reason it may be misleading to represent the associated physical hypothesis by P. A better representation would be "(P & G)," where G is the physical geometry used in the determination of areas and volumes that occur in P. Of course, use of euclidean geometry in the calculation of strains and stresses does not mean that G is euclidean as well.

Einstein concluded that it is possible to retain euclidean geometry in the face of any adverse empirical evidence simply by modifying the associated physical hypotheses. Of course, scientists may prefer a noneuclidean geometry in conjunction with physical hypotheses of simpler form, but this choice is not forced on them by the logic of the situation.

Grunbaum conceded that *if* deforming forces are present, then Einstein's account is correct. Application of the laws governing deformations (P) presupposes the selection of a geometry (g). Thus, where perturbations are present, the testing of the hypotheses of physical geometry (G) depends on antecedent geometrical assumptions (g).

Suppose that euclidean geometry is utilized in the application of laws governing deformations g_e. What adjustments in P would be necessary to retain the euclidean version of G in the face of *prima facie* negative evidence $\sim O$? G_e can be retained provided that a P^* can be formulated such that

$$(G_e \;\&\; P^* \;\&\; g_e) \to \sim O.$$

Grunbaum discussed two possibilities. The first possibility is that P can be changed to P^* by modifying the dependence of deformation on non-positional parameters such as temperature and magnetic field strength. The second possibility is that P can be changed to P^* only by introducing a new dependence of the deformation of rods on position and orientation.[36]

Grunbaum declared that the second possibility is far more likely. But to alter "(P & g_e)" in this way is to change the meaning of "congruence." The original semantic rules of the physical geometry G also would have to be changed to reflect this new dependence of congruence-determinations on position and momentum.

A move of this type would violate the condition of semantic stability that is necessary for the non-triviality of the Quine thesis. This is because the Quine thesis asserts, in this context, that given

$$[(G_e \& P \& g_e) \to O] \& \sim O,$$

1. there is a non-trivial P^* such that

$$(G_e \& P^* \& g_e) \to \sim O, \text{ and}$$

2. G_e is not altered in the process of replacing P with P^* (for instance, by changing the semantic rules of the theory). Grunbaum claimed to have shown by this argument that Einstein's version of the Quine thesis is unconvincing.

Grunbaum noted that if the *absence* of deforming forces could be established, then hypothesis G would be separately falsifiable, viz.,

$$([(G \& A) \to O] \& [\sim O \& A]) \to \sim G.$$

where A states the absence of perturbations in the region.

There is a technique to determine that no deforming forces are present within a region. This technique is to show that two solid rods of different chemical composition that coincide at one place also coincide everywhere else in the region, independently of their paths of transport. This coincidence is certifiable directly by observation and does not involve reference to any particular geometrical theory. Grunbaum cited this situation as a counter-case to the Quine thesis that it is only *collections* of hypotheses that are falsifiable.

Of course, one could argue that two rods that appear to differ chemically really have the same chemical composition. If this were the case, then the absence of deforming forces would not have been established. One could "save" geometrical hypothesis G by taking A to be false.

Grunbaum held that to pursue consistently such a strategy would be to deny that observation plays a role in the context of justification. If such rudimentary observations as those which establish differences in chemical composition are not to be trusted, then even total theoretical systems are immune from falsification. Given (H & A) → O and ~O, one would conclude, not that (H & A) is false, but that ~O is false. If this strategy were pursued consistently, then no observation statement would count against a given set of theoretical assumptions. According to Grunbaum, either the Quine thesis is false (since a counter-case has been identified), or observation is irrelevant to the justification of scientific theories. Since observation is relevant to the justification of theories, the Quine thesis is false.[37]

Scheffler's Defense of the Two-Tier View

In *Science and Subjectivity* (1967), Israel Scheffler reexamined the orthodox position in the light of criticisms that had been directed against it in the 1950s and early 1960s. Scheffler listed a number of theses basic to the orthodox view of the philosophy of science:[38]

1. There are observational and theoretical levels in the language of science.
2. The observational level contains empirical facts. An important aim of science is to explain these facts.
3. One way to explain facts is to deduce statements that record them from premises that include hypotheses.
4. Hypotheses are subject to intersubjectively repeatable tests in which their deductive consequences are compared with observation reports.
5. Theories explain laws by exhibiting them as deductive consequences of theoretical postulates.
6. Competing theories may be evaluated, in part, by reference to empirical facts in the domain in question.
7. Science is a cumulative enterprise in which experimental laws become incorporated into theories of progressively wider scope.

Scheffler thus included in his characterization of the orthodox position the two-tier view of theories, the deductive model of explanation, and the notion of progress through the reduction of theories.

Scheffler correctly emphasized that the critics' case against these orthodox dogmas rests on the claim that there can be no theory-neutral observation language. He conceded that it is necessary to abandon the position that there is some single observation language within which observations are recorded independently of all considerations of theory. No such language exists, and Hanson and Feyerabend are correct to insist that new theories create new ways of formulating observational evidence. But if there is no shared observational content between competing theories, then there can be no reduction of one theory to another.

The prospects for orthodoxy seemed bleak. Nevertheless, Scheffler maintained that a modified two-tier position is viable.

Critics of orthodoxy had stressed that the meaning of a term is a function of the language system in which it is embedded. Scheffler conceded that this is correct. But it also is true that terms with different connotative meanings in different theories may refer to the same objects. Feyerabend had discussed two theories in which color-predicates are ascribed to self-luminescent objects. In the first theory, color-predicate P_1 is interpreted to be independent of the relative velocity of observer and source. In the second theory, P_1 is interpreted to be a function of this relative velocity. Clearly, "P_1" has a different connotative meaning in each theory. But it well may be that it is one and the same set of self-luminescent objects that is referred to by each theory. Granted that P_1 is a property term in the first theory and a dyadic predicate in the second theory, it still may be the case that the same set of objects is named as bearers of properties in the first theory and *relata* in relative velocity expressions in the second theory. Scheffler noted that

> terms may denote the very same things though their synonymy relations are catalogued differently.[39]

In specific cases it is possible to determine the referents of terms without a prior determination of connotative meanings. For instance, the gas hydrogen may be identified by its "spectral fingerprint" independently of theoretical considerations about the internal structure of the hydrogen atom. Of course, reports about spectral lines are given in terms of the concepts of the electromagnetic theory of light—"wavelength," "frequency," et al. But these statements need not refer to the Bohr theory, the Bohr-Sommerfeld theory, or Heisenberg's version of quantum theory.

Scheffler maintained that sameness of reference provides a sufficient basis for

1. the relative stability of experimental laws (as opposed to theories);
2. the evaluation of competing theories by appeal to observation reports;
3. the possibility of the reduction of one theory to another; and
4. the cumulative nature of the scientific enterprise.

Scheffler's defense of a modified two-tier view was criticized by Hesse. In a review of *Science and Subjectivity*, she maintained that sameness of reference is neither sufficient nor necessary for the evaluation of theories.[40]

Sameness of reference is not sufficient because two theories may be about different aspects of the same set of objects. If the terms of T_1 refer to the shapes of the objects and the terms of T_2 refer to the colors of the objects, the two theories do not compete. The terms of the two theories may have the same denotative meaning without providing any basis for judging their relative merits.

Sameness of reference is not necessary for theory-evaluation either. For example, the denotation of the term "atom" was different in the theories of Dalton and Cannizzaro. One difference is that Cannizzaro, following Avogadro, did not predicate the term "atom" on the naturally occurring particles of common gases such as hydrogen, oxygen, and nitrogen. Despite this difference in the classification of objects, the two theories can be compared. For example, Cannizzaro's theory assigns the same atomic weights to each element for which both gas density data and the law of Dulong and Petit can be applied. Dalton's theory does not do this. Clearly Cannizzaro's theory is superior on observational grounds. Sameness of reference, then, is neither sufficient nor necessary for theory comparison.

Doubts about Explanation

Criticisms of the Covering-Law Model of Explanation
Subsumption under General Laws: A Necessary Condition of
Scientific Explanation?

Michael Scriven challenged the deductive-nomological model of scientific explanation in articles dating from 1959.[1] He maintained that subsumption under general laws is not a necessary condition of scientific explanation. Suppose that a bridge has collapsed following the explosion of a bomb nearby. According to Scriven, a perfectly adequate explanation of the collapse is "the bridge collapsed because a bomb exploded nearby."[2]

Scriven conceded that laws can be formulated that correlate explosive force, distance, and the tensile properties of bridge materials. But he argued that although these correlations may be cited to justify the explanation given above, they need not appear as premises of an argument that explains the collapse.

The orthodox theorist may reply that if an explanation is "justified" by appeal to a set of laws, then a deductive argument can be formulated in which these laws appear as premises. Hempel insisted that it is the corresponding deductive argument that provides the scientific explanation of the collapse.

Hempel pointed out, moreover, that a "justification" of "*q* because *p*" typically involves further particular statements as well as general laws. Scriven's distinction between "facts that explain" and "laws that justify explanations" must be amended. Statements about particular facts also may serve as grounds to justify an explanation. If an additional statement of fact P^* is mentioned in the Scrivenian explanation, then the explanation has been changed— "*q* because *p*" is replaced by "*q* because both *p* and P^*."

Scriven's claim that particular events often are explained without reference to general laws is not conclusive against the subsumption thesis. Scriven did not rest his case against the subsumption thesis on this point about scientific practice. He insisted that there are explanatory contexts in which

1. "*q* because *p*" fully explains *q*, and
2. no deductive-nomological argument can be formulated with *q* as its conclusion.

Consider the case of an ink-stained rug. A perfectly acceptable explanation of this tragedy is "the rug is stained because I knocked over the ink bottle."[3] This explanation makes an assertion of a causal connection. Scriven maintained that one may have adequate reasons for causal claims without being able to quote laws. A person may explain the accident mentioned above by confessing his clumsiness. It matters not that he cannot formulate the relevant laws of physics. Scriven emphasized that a caveman could have produced this explanation. The explanation has not become more certain with the formulation of the laws of inertia and elasticity.

If one were pressed to specify a true universal statement to use as a premise in order to deduce the conclusion in accordance with the deductive-nomological schema, it would have to be something like

> if you knock a table hard enough it will cause an ink-bottle that is not too securely or distantly or specially situated to spill (ink) over the edge (if it has enough ink in it).[4]

The particular premises would have to include a statement about knocking the table "hard enough" and a statement about the "relatively insecurely situated" bottle of ink. The required deductive relationship holds, but the law is very vague. We could eliminate phrases like "hard enough" from the law, and replace "will cause" by "probably will cause." However, on this interpretation the conclusion does not follow deductively from the law. Given "if *p* then probably *q*" and "*p*," it follows only that "*q* is probable." Scriven declared that the criterion of deduction must be abandoned if the criterion of universal hypotheses is abandoned; and what is left of the deductive model? We have instead an *inductive* model of explanation where for laws we have probability truisms and for deduction probability inference ... What is added to ... our [initial] explanation of the damaged carpet by production of some truism about the probable effect of knocking tables on which ink-bottles stand?[5]

Hempel complained that Scriven's ink-stained rug example gains plausibility from its vagueness. Scriven has not described in sufficient detail the phenomenon to be explained. The explanation has the form "q because p," but we are not told how to understand q. If q is a report that ink has leaked from an overturned, unstoppered ink bottle, then a deductive-nomological explanation is available in terms of the laws of fluid mechanics. However, if q is a report that includes the size and shape of the ink stain on the rug, then no laws are known from which q may be deduced. On this understanding of q, my knocking the table does not explain q either.

Scriven presumably had in mind a case in which the phenomenon to be explained is the production of an ink stain of unspecified size and shape. A wide variety of antecedent conditions may have produced the stain. Nevertheless, to select a particular set of antecedent conditions as the cause of a particular stain is to presuppose the applicability of covering laws.

Hempel held that to explain the last event in a sequence—table knocked, bottle overturns, ink drips onto rug, stain appears—one must do more than merely list the events. To declare "q because p" is to claim that the event described by q depends on the event described by p. And this is to claim that there is a covering law, which relates events similar to the event described by p and events similar to the event described by q.

Moreover, discovery of both covering laws and additional antecedent conditions is a task for scientific inquiry. Hempel maintained that Scriven was wrong to deny the relevance of scientific inquiry to the explanation of the stained rug. Scriven's caveman may have declared "q because p," but the adequacy of this claim depends on what it asserts. The adequacy of the caveman's "explanation" depends on the covering laws presupposed. For instance, the caveman may believe that every opaque liquid stains every floor covering. If his explanation presupposes this covering law it clearly is inadequate. By contrast, a contemporary scientist may seek to formulate a deductive-nomological explanation whose premises include well-confirmed laws of physics and chemistry. It may be that he cannot deduce q from a particular set of laws and antecedent conditions. However, this only shows the need for more detailed scientific inquiry. Hempel declared that the claim that the caveman could explain staining the rug with the same "certainty" as a modern scientist loses its initial striking plausibility when we ask ourselves precisely what the explanation would assert and what it would imply, and when we make sure it is not simply taken to be a narration of selected stages in the process concerned.[6]

Hempel's analysis is applicable as well to explanations of evolutionary change. Biologists sometimes advance "evolutionary histories" of the form "*q* because *p*." However, these histories gain explanatory import in virtue of implicit reference to covering laws.

Consider the case of the finches Darwin found on the Galapagos Islands. The state of affairs to be explained—*q*—is that a number of closely related species are distributed among the islands, such that a given species may be present on one or more, but not every, island.[7] An evolutionary history has been formulated to explain this state of affairs. In the distant past there occurred an initial dispersion of mainland finches to the islands. Once on the islands the ancestral finches encountered diverse living conditions. On some islands fruit was plentiful. On other islands seeds or particular types of insects were readily available. Once the initial migration had taken place, birds on a given island mated only with other birds on that island. Over time, separate distinct species emerged to fill the available environmental niches.

This evolutionary history has the form "*q* because *p*," but when the implicit generalization is stated the resultant argument can be expressed in deductive-nomological form, viz.,

1. $(x) [(Dx \,\&\, Gx \,\&\, Hx \,\&\, Tx \,\&\, Rx \,\&\, Ix) \supset Ox]$
2. $Da \,\&\, Ga \,\&\, Ha \,\&\, Ta \,\&\, Ra \,\&\, Ia$

$$\therefore Oa$$

where $Dx = x$ is a case in which there was an initial dispersion of mainland finches to the islands,

$Gx = x$ is a case in which geographic barriers sufficient to ensure reproductive isolation exist on each island,

$Hx = x$ is a case in which each island has a distinctive habitat,

$Tx = x$ is a case in which those finches in habitat H_1 that possess trait T^* are more suited to the performance of task K than are finches that lack T^*,

$Rx =$ x is a case in which success at task K affects positively its
          possessor's likelihood to survive and reproduce,

$Ix =$ x is a case in which T^* is transmitted genetically,

$Ox =$ x is a case in which individuals that posses T^* come to be
          dominant within H_1,

a is the specific case in which the long-beaked finch *geospiza
          scandens* has become dominant in habitat H_1 on James Island.

The first premise of this argument expresses a multiply conditional claim—*if*
there is an initial dispersion, *if* reproductive isolation ensues, *if* habitats differ, *if*
possession of a particular trait enables its possessor to succeed more readily at
some task, *if* success at that task affects positively the likelihood of survival and
reproduction, and *if* the trait is inherited, then evolutionary change results. The
argument achieves success only if each of the conditions is realized for the
specific case in question.

Multiply conditional arguments of the above type appear to be vulnerable to a
devastating criticism. This criticism is that the only way a biologist can establish
that the stated conditions are realized for a specific case is to appeal to the fact of
evolutionary change itself. It is only because closely related species of finches are
present on the Galapagos Islands that we know that an initial dispersion of
mainland finches took place, that reproductive isolation ensued, that the efficiency
of beaks of varying shapes in the acquisition of food affects reproductive chances,
and so forth. In short, the only way one can establish that premise (2) is true is to
appeal to the truth of the conclusion. But if there can be no independent evidence
for the truth of premise (2), then the explanation would appear to be circular.

Hempel insisted, however, that there need be no circularity present in "self-
evidencing" explanations. He noted that when a deductive-nomological
argument is used for explanatory purposes

> it does not claim to establish that E [the conclusion] is true; that is presupposed
> by the question "Why did the event described by E occur?"[8]

Information that the event described by E has occurred is not part of the
explanatory premises. If it were, then the explanation would be circular.

The explanatory process proceeds through the following stages:

1. assume E is true,

2. formulate an argument of deductive-nomological form with E as
   conclusion, and

3. show that the premises are true.

Stage (3) may include an appeal to the truth of *E* without introducing a circularity that defeats the explanatory process.

Of course, there would be an explanatory crisis if several deductive-nomological explanations of *E* were formulated, such that there was no evidence for the truth of any premise-set except the truth of *E*. However, the case of Darwin's finches is not so desperate. There is some independent evidence for the truth of the conjuncts of premise (2), and it is clear what sort of further evidence would be important.

Biologists have achieved extensive knowledge about adaptation. One conclusion they have drawn is that securing food is a task, success at which affects the likelihood of survival to reproductive age. In the case at hand, a long pointed beak is more efficient in obtaining nectar from a pear cactus plant. Given that the habitat of James Island features pear cactus plants, there is good evidence to support the claim that possession of a long pointed beak is an adaptive trait in that habitat.

Suppose, moreover, that new evidence is uncovered that the Galapagos Islands at one time were very close to the South American continent. This evidence might provide support for the initial dispersion claim. And the discovery of fossil remains on James Island judged intermediate in form between present species and the ancestral mainland species might support premise (2) as well. Evolutionary histories are not exceptions to Hempel's claim that subsumption under general laws is a necessary condition of scientific explanation.

The nomic subsumption thesis is subject to a further challenge, however. Wesley Salmon pointed out that one may formulate an effective causal explanation of an improbable event, even though neither an inductive-statistical nor a deductive-nomological explanation can be given for the event. Suppose Smith has developed leukemia following exposure to radiation. It may be the case that only 1 percent of those similarly exposed develop the disease. Nevertheless one can explain Smith's illness by citing his exposure to radiation. A full explanation would refer to such causal processes as the production of gamma rays upon nuclear fission, the modification of cellular structure by gamma rays, and the differential response of modified and unmodified cells to attack by the leukemia virus. Salmon emphasized that what counts in an explanation of this type is the statistical relevance of the contrast between exposure and nonexposure and the subsequent development of leukemia.[9]

The radiation-leukemia case does not fit the inductive-statistical pattern because the value of the statistical correlation is much less than 1.0. This case does not fit the deductive-nomological pattern either. A deductive-nomological explanation can be given for the low probability of the occurrence of Smith's illness, but not for the illness itself. Thus there are events for which satisfactory scientific explanations can be given that fit neither the deductive-nomological pattern nor the inductive-statistical pattern. The nomic subsumption requirement is too strong. Instantiation of one of the two patterns is not a necessary condition of scientific explanation.

Subsumption under General Laws: A Sufficient Condition of Scientific Explanation?

Is subsumption under general laws a sufficient condition for scientific explanation? In particular, does a set of statements qualify as a scientific explanation just because it satisfies the Hempel and Oppenheim schema for deductive-nomological explanation?

Consider the following exchange: "Why is that flame green? Because it is a barium-affected flame, and all barium-affected flames are green". The answer satisfies the deductive-nomological schema, but it merely generalizes the initial puzzlement. William Dray insisted that

> when puzzled by something, we do not ordinarily find it enlightening to be told: "That's what always happens."[10]

Rom Harre pointed out that an appropriate explanation of the characteristic color of barium-affected flames involves an appeal to such topics as ionization, molecular orbital theory, and quantum theory.[11] According to Harre, we should distinguish two sorts of explanation of an event: a "minimal explanation" in which the event is subsumed under a general law, and a "scientific explanation" in which the event is represented as the outcome of the operation of some underlying mechanism.[12]

At the very least, it would seem that a defender of orthodoxy would have to rank a deductive-nomological explanation in terms of atomic theory higher than a deductive-nomological explanation in terms of a correlation of flame color and the presence of barium. But, so far as I am aware, no philosopher of science has claimed that one deductive-nomological explanation is as good as any other. Moreover, Hempel, for one, denied that every set of statements that satisfies the deductive-nomological schema qualifies as an acceptable scientific explanation. He called attention to the following example suggested by S. Bromberger:

Theorems of physical geometry.
Flagpole *F* stands vertically on level ground
and subtends an angle of 45 degrees when
viewed from ground level at a distance of 80 feet.

∴ Flagpole *F* is 80 feet in length.

Hempel emphasized that the premises of this argument do not explain why the flagpole is 80 feet in length.[13]

Hempel also called attention to the two deductive-nomological arguments listed below:

$$1.\ T \propto \sqrt{l} \qquad\qquad 2.\ T \propto \sqrt{l}$$
$$\frac{l_2 = 1/4\, l_1}{\therefore T_2 = 1/2\, T_1} \qquad\qquad \frac{T_2 = 1/2\, T_1}{\therefore l_2 = 1/4\, l_1}$$

where *T* is the period of a pendulum and *l* is its length. We readily accept argument (1) as an explanation of the change in period of the pendulum. However, we are hesitant to accept argument (2) as an explanation of the change of length of the pendulum. Presumably this is because we can saw through a pendulum and thereby alter its period, but we cannot alter its period independently of changing its length. Hempel noted, however, that period and length are symmetrically related. It is true that we cannot alter its period without changing its length. But it also is true that we cannot change its length without altering its period. Hempel declared that

> in cases such as this, the common sense conception of explanation appears to provide no clear and reasonably defensible grounds on which to decide whether a given argument that deductively subsumes an occurrence under laws is to qualify as an explanation.[14]

Hempel thus did not maintain that deductive-nomological subsumption is a sufficient condition of scientific explanation.

Criticisms of the Alleged Explanation–Prediction Symmetry

Another point of vulnerability of the orthodox position was the alleged symmetry of scientific explanation and scientific prediction. Hempel and Oppenheim had tossed off, almost as an aside, a suggestion that explanation and prediction are symmetrical relations. They pointed out that both explanation and prediction may satisfy the deductive-nomological schema. In their view, the essential difference between the two is one of direction. In explanation one begins with a conclusion and selects a set of premises from which it may be deduced; in prediction one selects premises and derives a conclusion.[15]

Although Hempel and Oppenheim did not elaborate on the symmetry thesis, critics assumed that the thesis involved two claims:

1. Every successful deductive-nomological prediction of an event counts as an explanation of why the event occurred.
2. Every satisfactory deductive-nomological explanation of an event is potentially a prediction of the occurrence of that event. That is to say, if the laws and antecedent conditions had been known prior to the event, then the deductive-nomological schema could have been applied to predict that the event would occur.

The first claim is false. Several writers have emphasized that for 2000 years scientists have distinguished between "saving the appearances" and explaining why appearances are as they are.[16] This distinction was implicit in Greek and Medieval astronomy. Astronomers created mathematical models to predict the positions of the planets. The epicycle-deferent model was widely used for this purpose.

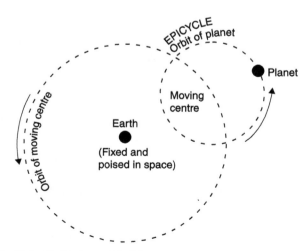

Figure 24 The Epicycle-Deferent Model.

Using this model, one could predict, for example, the position of Venus against the background stars. However, one could not explain why Venus was at a particular location by formulating a deductive argument, the premises of which include statements about epicycles and deferents. No claim was made that the "motions" in the model were of the same kind as the real motions of the planets.

In a post-"Hempel and Oppenheim" article, Hempel conceded that arguments effective for purposes of prediction may not have explanatory force. This often is the case in medical prognoses. Hempel gave the following example:

In all cases, appearance of Koplik spots on the inside of
the cheeks at time t_1 is followed by measles at t_2.
Subject S has Koplik spots at t_1.

$\therefore S$ has measles at t_2.[17]

Hempel admitted that such arguments do not have explanatory force.

Nevertheless, he continued to maintain that the converse claim is true. Every satisfactory explanation of the occurrence of an event is potentially a prediction of that event.

A number of critics rejected this claim. Nicholas Rescher complained that questions about a symmetry between explanation and prediction are empirical questions.[18] Whether or not one can use information about the present state of a physical system to predict a future state depends on the laws that govern the system. For some systems, a satisfactory deductive-nomological explanation of state (2) in terms of antecedent state (1) could have been used at the time of state (1) to predict the occurrence of state (2) at a later time. Inferences to earlier states also can be made. For other systems, the appropriate laws are not temporally symmetrical. In the case of heat flow across a boundary, for example, one can deduce, from information about the present state, information about future states, but not information about prior states. And in the case of goal-directed servomechanisms, it sometimes is possible to make inferences about earlier states without being able to make corresponding inferences about future states.

N. R. Hanson pointed out that the explanation–prediction symmetry holds for Newtonian dynamics but not for quantum physics. Any planetary motion explained by reference to Newtonian formulas could have been predicted from knowledge of an earlier state of the system. The case is different for the emission of α-particles. After the fact of emission, the equations of quantum theory may be used in a deductive-nomological argument to explain the event. However, one cannot use the theory to predict that an α-particle will be emitted at some specific future time.

Hanson conceded that one could hold that our inability to predict an individual micro-event counts against the claim that we have explained it. But what physicists count as "explaining" a micro-event is to deduce a statement describing it from premises that include the laws of quantum physics. Hanson declared that

> the philosopher should not legislate here; he must note what *counts* as explanation in micro-physics and then describe it precisely.[19]

From this standpoint, Hanson concluded that the symmetry thesis is false.

Israel Scheffler also advanced arguments against the alleged explanation–prediction symmetry.[20] The symmetry thesis fails, according to Scheffler, because of temporal asymmetries not taken into account by Hempel and Oppenheim. In particular, a post-dictive inference may fulfill the requirements of the deductive-nomological explanation without constituting a potential prediction of the event explained. Consider the deductive-nomological argument:

$$\frac{A \;\&\; L}{\therefore B}$$

where *A* is a statement describing event *a*, and *B* is a statement describing event *b*. Let *A* describe the present configuration of sun, moon, and earth, and *B* describe a past eclipse of the sun. A contemporary astronomer may use the laws of Newtonian mechanics to deduce *B* from *A*. This is a *bona fide* case of explanation. But no question about prediction can arise. For a predictive argument *a* must precede *b*.

Scheffler pointed out that if one accepts Hempel's claim that "an explanation of a particular event is not fully adequate unless its explanans, if taken account of in time, could have served as a basis for predicting the event in question," then the argument "A & L / B" would not qualify as an explanatory argument. This is because "(A & L)" could not be used to predict *b*. Scheffler suggested that since the "(A & L /∴ B)" argument meets the requirements that Hempel and Oppenheim placed on deductive-nomological explanations, an additional requirement is needed. This requirement is that *a* must not be prior in time to *b*.

Michael Scriven came to assume leadership of the anti-symmetry forces. In a series of papers on explanation, Scriven created various counter-cases to the claim that every adequate deductive-nomological explanation is a potential prediction of its conclusion.[21]

One such case is the collapse of a bridge due to metal fatigue. Given the fact of collapse, a sound deductive argument may be formulated as follows:

$$(x) [(Lx \lor Dx \lor Fx) \supset Cx]$$
$$Fa$$

$$\therefore Ca$$

where $Lx = x$ is an instance of excessive load,

$Dx = x$ is an instance of external damage,

$Fx = x$ is an instance of metal fatigue, and

$Cx = x$ is an instance of bridge collapse.

Scriven argued that this argument does not have predictive force, because the truth of the premise stating that metal fatigue is present can be established only after the fact of collapse. That is to say, whether fatigue sufficient to cause collapse of the bridge is present can be ascertained only in terms of collapse or non-collapse itself.[22]

A second counter-case suggested by Scriven involves the connection between syphilis and that form of paralysis known as paresis. Very few persons who have syphilis develop paresis, but the only known cause of paresis is syphilis. Scriven maintained that

"The only known cause of paresis is syphilis

P has syphilis

$\therefore$ *P* develops paresis"

is a legitimate explanatory argument.[23] However, it would not be an adequate argument for the purposes of prediction, since very few syphilitic persons develop paresis. What can be predicted successfully from knowledge of this connection is only that the chance that a person develop paresis increases slightly (from zero) if he has syphilis.

Further counter-cases suggested by Scriven involve applications of the theory of organic evolution.[24] He noted that we sometimes can explain why a certain species survived changes in environmental conditions even when we could not have predicted its survival. For instance, we sometimes can explain the survival of animals equipped to swim in the case of the flooding of their natural habitat, without being able to predict this survival. Scriven acknowledged that "hypothetical probability predictions" could be made in such cases, viz., "if a flood occurs, then these animals would be likely to survive." He maintained that such "predictions" have value for actual predictions only if we can know independently of the event in question that the relevant conditions are realized. In many applications of evolutionary theory this knowledge is not available.

A defense of the symmetry thesis against these objections was given by Adolf Grunbaum in an essay published in 1963,[25] Grunbaum presented the following reconstruction of Hempel's position.

In explanation, the phenomenon described in the conclusion (*E*), *belongs to a scientist's past at the time he accounts for it.* "Hempelian-explanation," thus construed, includes both prediction and retrodiction.

In prediction, *E belongs to a scientist's future at the time he accounts for it.* "Hempelian-prediction," thus construed, also includes both prediction and retrodiction. The antecedent conditions cited in the premises of a "Hempelian-prediction" for which *E* is later than the antecedent conditions may be either in the scientist's past or his future.

Grunbaum maintained that the logical relation between explanation and prediction is the same for Hempelian-explanation and Hempelian-prediction. This is the relevant symmetry between explanation and prediction. A genuine counter-case against this symmetry thesis would be one for which the logical relation between event explained and explanatory premises is stronger (or weaker) than the corresponding argument used to explain the event.

Grunbaum noted that critics such as Rescher, Hanson, and Scriven have focused instead on a scientist's ability to draw conclusions from a body of evidence. Rescher, for example, correctly maintained that, for certain systems, a state-description from which information about a future state can be predicted cannot be used to retrodict information about an earlier state. And Hanson correctly maintained that, in quantum mechanics, reliable knowledge that a specific kind of micro-event has occurred has no counterpart in our knowledge of the future occurrence of such an event. However, the relevant symmetry is the logical relation between premises and conclusion in each case. Grunbaum argued that, in quantum mechanics, this logical relation is irreducibly statistical for both Hempelian-explanation and Hempelian-prediction.

Grunbaum dealt with Scriven's paresis counter-case in much the same way. Granted that syphilis (*S*) is the only known cause of paresis (*P*), the following translations are appropriate:

1. "*S* did cause *K* to have *P*" = "*K* has (or had) *P* and it was caused by *S*," and
2. "*S* will cause *Z* to have *P*" = "*Z* will have *P* and it will have been caused by *S*."[26]

Grunbaum conceded that (1) and (2) are asymmetrical with respect to the *assertability* of the explanandum. That K has paresis is a matter of record; that Z will have paresis is not a matter of record. But the *inferability* of paresis from syphilis is symmetrical with respect to time. Grunbaum declared that

> insofar as a *past* occurrence of paresis can be inductively inferred from prior syphilis, so also a future occurrence of paresis can be. For the causal relation or connection between syphilis and paresis is incontestably time-symmetric: precisely in the way and to the extent that syphilis *was* a necessary condition for paresis, it also *will* be.[27]

At this point in the debate, Scriven cried "foul." He pointed out that initially the orthodox (Hempelian) position was that there is an important symmetry between explanation and prediction. Critics of the symmetry thesis advanced a number of counter-cases. And now Grunbaum sought to render ineffective the counter-cases by suggesting that the symmetry thesis was intended to apply only to "Hempelian-explanation" and "Hempelian-prediction." This is unusual usage. "Hempelian-predictions" include both predictions and retrodictions. Grunbaum has sought to assimilate understanding and inferability.[28]

Scriven emphasized that deductive subsumption often does not constitute explanation. For instance, one may infer the occurrence of a storm from a sudden drop in the reading of a barometer and an "indicator law" that correlates such readings with storms. However, one does not explain the storm by citing the barometer readings and the indicator law.

On this point Hempel proved more flexible that Grunbaum. Hempel conceded that some arguments effective for the purpose of prediction do not have explanatory force. Grunbaum insisted, on the other hand, that a deductive-nomological argument that employs an indicator law does provide a measure of explanation. He recognized two grades of deductive-nomological explanation in science: (1) subsumption under a causal law, and (2) subsumption under an indicator law. Scriven took Grunbaum's introduction of grades of explanation to be another reason to reject the identification of scientific explanation and inferability.

Hempel continued to insist that every adequate explanation of the occurrence of an event is potentially a prediction of that event. He maintained that any rationally acceptable answer to the question "why did event E occur?" must include good grounds for believing that E did in fact occur. According to Hempel, *if* an adequate deductive-nomological explanation of the occurrence of an event is available, and *if* the information stated in the premises is known and taken into account, *then* a prediction of the event could have been made. The counter-cases proposed by the critics are ineffective against this claim.

For example, in the metal-fatigue-bridge-collapse argument, it is true that we do not know that the conditions specified in the explanans are true independently of our knowledge that the event described in the explanandum took place. This leaves untouched the conditional thesis that *if* we had information about the presence of a certain degree of metal fatigue *then* we could have predicted the collapse of the bridge.[29]

The syphilis-paresis case also is ineffective against the conditional claim, but for different reasons. According to Hempel, the purported "explanation" of paresis is not satisfactory. Because paresis seldom follows prior syphilitic infection, one cannot explain a case of paresis by citing a prior case of syphilis. To explain an event, it does not suffice merely to specify a necessary condition of its occurrence. If it did, then one could explain why a man won the Irish sweepstakes by pointing out that he had purchased a ticket, and that only those who have purchased tickets are eligible to win.[30]

In a review of the literature on the paresis example, Paul Dietl concluded that neither Grunbaum nor Hempel had provided a satisfactory analysis.[31] Scriven had maintained that

1. we can explain a present case of paresis by reference to its only known necessary condition—syphilis—and
2. we cannot predict that an individual who has syphilis is likely to develop paresis (although we do know that he is more likely to develop paresis than he would be if he did not have syphilis).

Dietl observed that Grunbaum's distinction between "did cause" and "will cause" is beside the point. Grunbaum had suggested the following translation: "*S* will cause *Z* to have *P*" = "*Z* will have *P* and it will have been caused by *S*."

Assume that *Z* does have syphilis. Grunbaum has conceded that *Z*'s contracting paresis is not a matter of record. But then Grunbaum's translation merely states that *if Z* develops paresis then its cause is the prior condition of syphilis. This is a conditional prediction that does not invalidate Scriven's claim that we cannot predict that *Z* is likely to develop paresis.

Dietl also maintained that Hempel had missed the point of the paresis example. Syphilis is a factually necessary condition of paresis. But the purchase of a ticket is a logically necessary condition of winning the Irish sweepstakes. Of course, one might gain the prize by stealing or counterfeiting a ticket, but this would not count as "winning" the sweepstakes. Hempel's "explanation" of winning the sweepstakes is analogous to the following "explanation" of Jones" divorce:

Jones married his wife.

Only married persons can sue for divorce.

∴ Jones divorced his wife.

In a subsequent essay, Hempel emphasized that, although syphilis is a factually necessary condition of paresis, the paresis-syphilis argument is too incomplete to qualify as an explanation.[32] He maintained that an adequate explanation of the incidence of paresis would have to include additional information. If the additional conditions whose presence makes the difference between a syphilitic person's developing or not developing paresis are known and taken into account, then paresis could be predicted.

Scriven acknowledged that the syphilis-paresis explanation is incomplete. We cannot specify in detail the conditions that make the difference between developing and not developing paresis. One day it may be possible to predict which syphilitic persons will develop paresis. But

> until that day where every thing is predictable, there remains the fact that we can often explain what we could not predict, and surely this feature should be mirrored in any analysis of these notions.[33]

Hempel's conditional thesis is not of much interest. Scientific explanations normally are incomplete. Scriven insisted that the fact that a given explanation is incomplete ought not disqualify it from being an explanation.

New Problems about Confirmation and Justification

Goodman's "New Riddle of Induction"

It was the orthodox position that a universal generalization—"$(x) (Ax \supset Bx)$"—is confirmed by its positive instances—"Aa & Ba," "Ab & Bb," ... A black raven, for instance, confirms the generalization that "All ravens are black." In an important study published in 1953, Nelson Goodman argued that the principle of instance confirmation should be rejected.[1] He pointed out that some universal generalizations do not receive support from their positive instances. A case in point is the generalization

1. "All emeralds are grue,"

where "x is grue" if, and only if, "either x is examined before time t and is green, or x is not examined before time t and is blue."[2]

Emeralds examined before time t support both (1) and generalization (2) below:

2. "All emeralds are green."

Suppose that t is 12 p.m. tonight. Which generalization shall we rely on to predict the color of emeralds examined tomorrow? The evidence at hand supports equally (1) and (2). It also supports generalizations about emeralds that are "grellow," "grurple," "gred," etc. Our intuitions are offended. We "know" that instances of green emeralds provide support for (2) and not for (1). And yet every instance that confirms (2) also confirms (1).

Goodman noted that neither the generalization about grue emeralds nor accidental generalizations like "All the coins now in my pocket contain copper" receive support from their positive instances. That a particular coin now in my pocket contains copper provides no support for the claim that another coin now in my pocket also contains copper. The relationship is different for nomological universals. Evidence that one sample of sodium reacted with chlorine does support the claim that another sample of sodium also reacts with chlorine.

"Grue" is a strikingly artificial and complex predicate. If some way could be found to exclude such predicates from the language of science, generalizations like (1) could not be formed. Perhaps we might

1. restrict instance confirmation to generalizations about "basic predicates";
2. take "basic predicates" to be those predicates that cannot be defined in terms of other predicates; and
3. select "green" and "blue" as "basic predicates."

"Grue" then would be a non-basic predicate to which the principle of instance confirmation does not apply.

Goodman argued that an approach of this type is ineffective. It is ineffective because the relationship between the pair ["green," "blue"] and the pair ["grue," "bleen"] is symmetrical. ("x is bleen" if, and only if, "either x is examined before time t and is blue, or x is not examined before time t and is green"). The problem is that we also might select "grue" and "bleen" to be "basic predicates," in which case "green" and "blue" would be non-basic predicates, e.g., "x is green" if, and only if, "either x is examined before time t and is grue, or x is not examined before time t and is bleen."[3]

"Grue" does have one feature that "green" lacks—explicit reference to a particular time. This suggests that it might be helpful to subdivide predicates into those that make reference to temporal or spatial values, and those that do not. The principle of instance confirmation then could be restricted to generalizations about "purely qualitative" predicates that involve no reference to time or space.

Goodman maintained that this approach will not work either. He noted that the riddle about emeralds can be restated without reference to time.[4] Since there has been a finite set of emeralds that have been examined and found to be green, the predicate "grue" may be defined with respect to this finite set, viz., "x is grue" if, and only if, "x is identical with $(a \lor b \lor c \lor \ldots n)$ and is green or $\underline{x}$ is not identical with $(a \lor b \lor c \lor \ldots n)$ and is blue."

On this definition of "grue," application of the principle of instance confirmation accords equal status to both (1) and (2). The riddle remains.

Goodman's own approach to the riddle is to refer to the past usage of predicates. Predicates such as "green" and "blue" occur in generalizations that have been used to account for new instances. "All emeralds are green" and "All barium-affected flames are green" are generalizations that have been applied in this way. "Grue" and "bleen" do not occur in generalizations with a similar history of usage in applications to new instances. Goodman suggested that predicates be classified with respect to their "track record" of previous usage. Certain predicates qualify as "entrenched predicates"[5] to which the principle of instance confirmation is relevant. Entrenched predicates have participated in generalizations that have been projected to account for additional instances. "Grue" is not an entrenched predicate. Of course, scientists might have projected "All barium-affected flames are grue" onto additional instances. But as a matter of fact they did not. What counts is actual usage. The biographies of "green" and "grue" are markedly different.

Given that some generalizations have been projected onto additional instances and others have not, Goodman sought to identify the distinguishing features of projectable generalizations. According to Goodman, a projectable, and hence lawlike, generalization G satisfies a set of requirements that include the following:

1. G is supported by evidence from its positive instances.
2. There are no negative (disconfirming) instances.
3. There are additional instances over and above those in the evidence class.

4. There is no conflict in projection with a generalization *H* such that *H* has a no-less-well-entrenched antecedent-predicate and a better-entrenched consequent-predicate.[6]

This is not the complete set of requirements that Goodman elaborated. But it is sufficient to indicate how accidental generalizations like (2) and (3) can be excluded from the class of projectable generalizations. "All emeralds are grue" is excluded upon application of requirement (4). "All coins in my pocket contain copper" is excluded upon application of requirement (3).

Goodman's requirements must be applied to the historical record of the various predicates that occur in generalizations. For this reason, every determination of the projectability of a generalization is relative to a particular point in time (and perhaps also to a particular location). For example, given the available evidence class, "All swans are white" was a projectable generalization in 1750, but ceased to be such upon discovery of black swans in Australia.

One effect of Goodman's discussion was to question the value of studying confirmation in artificial languages. Hempel's satisfaction theory, for instance, takes confirmation to be a syntactical relationship between a hypothesis and an evidence statement. Confirmation is defined in terms of the relation of logical entailment. A presupposition of this theory of confirmation is that the language within which the confirmation-relation holds contains a clearly delimited set of observation-predicates. Goodman has shown that unless each of the predicates of this language is well-entrenched, some hypotheses of the form "(x) (Ax ⊃ Bx)" can be formed that receive no confirmation from their positive instances. In a "Postscript" (1964) to his 1945 essay, Hempel conceded that

> the search for purely syntactical criteria of qualitative or quantitative confirmation presupposes that the hypotheses in question are formulated in terms that permit projection; and such terms cannot be singled out by syntactical means alone.[7]

Historical Theories of Confirmation

The orthodox approach to confirmation had been to search for a purely logical relation between a hypothesis and sentences reporting evidence. The logical reconstructionist position was that a universal conditional—"(x) (Ax ⊃ Bx)"—is confirmed by its positive instances "Aa & Ba," "Ab & Bb," ... It does not matter whether "Aa & Ab" is affirmed to be true before, or after, hypothesis "(x) (Ax ⊃ Bx)" is formulated.

After two decades of debate about "black ravens" and "grue emeralds," the orthodox approach had lost its appeal. Hempel had conceded that the projectability of predicates cannot be determined on syntactic grounds alone.[8] If this is the case, then prospects are not good for a purely logical theory of the confirmation of hypotheses.

Popper, Kuhn, and critics of logical reconstructionist philosophy of science emphasized that the temporal relationship between hypothesis and evidence is of great importance. Popper, in particular, insisted that it is not instances, per se, that provide evidential support for hypotheses, but tests.[9]

The temporal symmetry of hypotheses and instances is broken if confirming evidence is restricted to tests. One cannot test a hypothesis before it is formulated. On Popper's position, there is a historical dimension to the confirmation of hypotheses.

Scientists sometimes do attach great significance to dramatic successful predictions. The work of Dmitri Mendeleeff is a good illustration. Mendeleeff hypothesized that there is a periodic dependence of the chemical properties of the elements on their atomic weights. He based this hypothesis on knowledge of the properties and atomic weights of some known elements (see Figure 25).

The sequence in Figure 25 was part of the database the hypothesis was designed to explain. The hypothesis was created precisely to account for these regularities.

When Mendeleeff extended the sequence to elements of higher atomic weights, he noted that there were gaps in the table. He attributed the gaps to yet-to-be-discovered elements. Mendeleeff applied the Periodic Law to predict the physical and chemical properties of the missing elements (labeled "eka-aluminum," "eka-boron," and "eka-silicon"). The missing elements subsequently were discovered and Mendeleeff's predictions proved to be remarkably accurate.

These predictions were successful *tests* of the Periodic Law. On a historical theory of confirmation, these "new" data provide important evidential support. The antecedently known correlations between chemical properties and atomic weights do not.

element	Li	Be	B	C	N	O	F
atomic weight	7	9	10	12	14	16	19
	Na	Mg	Al	Si	P	S	Cl
	23	24	27	28	31	32	35.5

Figure 25 Mendeleeff on the Periodic Properties of Chemical Elements.

	Prediction	Determination
	eka-Aluminium	Gallium
atomic weight	68	69.9
atomic volume	11.5	11.7
	eka-Boron	Scandium
atomic weight	44	43.79
oxide	Eb_2O_3	Sc_2O_3
	eka-Silicon	Germanium
atomic weight	72	72.3
oxide	EsO_2	GeO_2

Figure 26 Mendeleeff's Predictions (1871).

Source: J. W. van Spronsen, *The Periodic System of Chemical Elements* (New York: Elsevier, 1969), 139.

A purely temporal theory of confirmation would disallow certain obvious instances of evidential support. One such instance is Galileo's theory of projectile motion. Gunners were aware that the maximum range of a projectile is achieved at an angle of 45 degrees to the ground. Subsequently, Galileo introduced the hypothesis that projectiles fired from guns describe parabolic trajectories through the air. He deduced that the maximum range is achieved at an angle of 45 degrees. Clearly the prior data accumulated by gunners should count as providing support for Galileo's hypothesis.

A second instance is Einstein's application of the hypothesis of energy quantization in his theory of the photoelectric effect (1905). Prior to Einstein's work, Max Planck had appealed to the hypothesis of energy quantization to explain black-body radiation (1900). The observational evidence from black-body radiation provides support for Einstein's theory. It would be overly restrictive to disqualify the black-body data just because it was obtained before Einstein formulated his theory about the photoelectric effect.

Elie Zahar responded to these obvious shortcomings of a purely temporal theory of confirmation. He suggested that facts not taken into account in the formulation of a hypothesis may count as providing evidential support for the hypothesis.[10] This is the case even if these facts were established before the hypothesis was entertained. Even if Galileo developed his theory of projectile motion from geometrical analysis and experiments on inclined planes, without considering the testimony of gunners, the pre-theoretical testimony of gunners provides support for the theory of parabolic projectile trajectory.

A much discussed example that supports Zahar's position is the relation between Einstein's theory of special relativity and the null result of the Michelson-Morley experiment. Michelson and Morley provided evidence that the velocity of light is the same both in the direction of the earth's orbital motion and at right angles to this motion. Since this result was known before Einstein developed his theory, it would not support that theory on a purely temporal view of confirmation. Moreover, Einstein himself insisted that the Michelson-Morley result did not play a role in the formulation of special relativity theory.[11] But clearly the Michelson-Morley experiment is evidence that supports the theory. One cannot restrict evidential support to experimental results actually taken into account by the theorist when he formulated the theory.

Nevertheless, it may be objected that it is wrong to base decisions about evidential support on the biography of the theorist. Alan Musgrave, among others, objected to the subjectivist emphasis of Zahar's approach to confirmation.[12]

Imre Lakatos advanced an objective, historical theory of confirmation.[13] He recommended a three-term relation of evidential support. This relation links an evidence statement e to both hypothesis H and a "touchstone hypothesis" H_T. A "touchstone hypothesis" is a hypothesis in competition with H. Lakatos recommended the following criterion of evidential support:[14]

> e supports H only if
>
> 1. H, in conjunction with statements of initial conditions and boundary conditions, implies e, and either
> 2a. H_T, in conjunction with the same statements, implies that e is not the case (even if e was known prior to H), or
> 2b. H_T, in conjunction with the same statements, implies neither e nor $\sim e$.

In each of the examples in Figure 27 from the history of science, e supports H on Lakatos' criterion.

"Degree of Confirmation" and the Justification of Hypotheses

Goodman's work placed a roadblock in the path of the development of a purely syntactical definition of qualitative confirmation. One response of confirmation-theorists was increased interest in theories of quantitative confirmation. However, difficulties arose for orthodoxy in this area as well.

e	H	H$_T$
weight calx > weight corresponding metal	Oxygen Theory	Phlogiston Theory
no pores found in septum of heart	Harvey's Theory of Circulation	Galen's Theory of Ebb and Flow
c $_{air}$ > c $_{water}$	Maxwell's Electromagnetic Theory	Newtonian Corpuscular Theory
Michelson-Morley experiment	Special Relativity Theory	Aether Theory
Anomalous Perihelion of Mercury	General Relativity Theory	Newtonian Mechanics

Figure 27 Hypotheses, Touchstone Hypotheses and Evidence Statements.

In papers published in the 1950s, Karl Popper addressed the problem of quantitative confirmation. He noted that an adequate measure of quantitative confirmation would be a measure of the growth of scientific knowledge. To give high marks to hypothesis *H* is to indicate its superiority to earlier hypotheses about the same subject matter.

Popper presented arguments to support both a positive thesis and a negative thesis about quantitative confirmation. The positive thesis is that a definition of "degree of corroboration" should satisfy certain requirements. The negative thesis is that Carnap's concept "degree of confirmation" is not a suitable interpretation of the probability calculus.

Popper's Negative Thesis

Popper sought to show that it is inappropriate to equate "the degree to which statement *y* confirms statement *x*" and "the relative logical probability of *x*, given *y*." His approach was to show that there are cases in which a given statement both "confirms" a second statement and decreases its relative logical probability. Popper's argument is reproduced below:[15]

If statements $x_1, x_2,$ and y can be found such that

(a) x_1 and x_2 are independent of y, and
(b) y supports the conjunction $(x_1 \& x_2)$,

then y confirms $(x_1 \, \& \, x_2)$ to a higher degree than y confirms x_1, viz.,

(1) $C\,(x_1, y) < C\,[(x_1 \, \& \, x_2), y]$

where "$C\,(x_1, y)$" stands for "the degree to which y confirms x_1."

Line (1) is incompatible with the view that "degree of confirmation" is a probability, because line (2) below is a consequence of the axioms of the calculus of probabilities.

(2) $P\,(x_1, y) \geq P\,[(x_1 \, \& \, x_2)\,y]$

where "$P\,(x_1, y)$" stands for "the probability of x_1, given y."

Consider a set of colored counters—a, b, c, ...—each having one of four mutually exclusive and equally probable colors—red, blue, yellow, or green. The following probabilities may be derived:

Sentence	Probability
x_1—"a is blue or green"	$P\,(x_1) = 1/2$
x_2—"a is blue or red"	$P\,(x_2) = 1/2$
y —"a is blue or yellow"	$P\,(y) = 1/2$
x_1, given y	$P\,(x_1, y) = 1/4$
x_2, given y	$P\,(x_2, y) = 1/4$
$(x_1 \, \& \, x_2)$, given y	$P\,[(x_1 \, \& \, x_2)\,y] = 1/2$

The result—$[P\,(x_1, y) = 1/4] < [P\,(x_1 \, \& \, x_2), y] = 1/2$—contradicts theorem (2), which states

(2) $P\,(x_1 \, y) \geq P\,[(x_1 \, \& \, x_2), y]$.

Popper concluded that the axioms of the probability calculus do not provide an appropriate basis to explicate the concept "degree of confirmation."

He then developed an even more striking example to show that the logical probability of a sentence cannot be taken as a measure of its degree of confirmation. Given the set of colored counters, the following probabilities hold:

Sentence	Probability
x_1—"a is blue"	$P\,(x_1) = 1/4$
x_2—"a is not red"	$P\,(x_2) = 3/4$
y—"a is not yellow"	$P\,(y) = 3/4$
x_1, given y	$P\,(x_1, y) = 1/3$
x_2, given y	$P\,(x_2, y) = 2/3$

Clearly *y supports x₁*, since $P(x_1) = 1/4$ and $P(x_1, y) = 1/3$.

Moreover, *y undermines x₂*, since $P(x_2) = 3/4$ and $P(x_2, y) = 2/3$.

But if y supports x_1, and y undermines x_2, then y confirms x_1 to a greater degree than y confirms x_2. No matter how we choose to define "degree of confirmation"—C—we demand, for the above case, that

$$C(x_1, y) > C(x_2, y).$$

Popper concluded that since $P(x_1, y) < P(x_2, y)$ for this case, it would be inappropriate to take logical probability as a measure of degree of confirmation.

The two cases above support Popper's negative thesis. If the relative logical probability of a statement is taken as a measure of its degree of confirmation, counterintuitive consequences result. Since the original purpose of interpreting confirmation-functions to be probabilities was to provide a quantitative measure of the strength of hypotheses relative to a body of evidence, a different approach to the measurement of "degree of confirmation" is indicated.

Replies on Behalf of Carnap

There is little doubt that Popper's negative thesis was directed at Rudolf Carnap. In a footnote of the 1954 paper, Popper maintained that Carnap had identified "degree of confirmation of *x*, given *y*" and "relative logical probability of *x*, given *y*."

John G. Kemeny suggested that, contrary to what Popper claimed, Popper and Carnap had been discussing different concepts. According to Kemeny, Carnap's "degree of confirmation" is a measure of "how sure we are of *x* if we are given *y* as evidence," whereas Popper's "degree of confirmation" is a measure of "how much surer we are of *x*, given *y*, than without *y*."[16] This being the case, Popper's negative thesis does not engage Carnap's work.

Y. Bar-Hillel also rose to the defense of Carnap.[17] Bar-Hillel maintained that Popper was guilty of confusing Carnap's usage of "degree of confirmation" with a "pre-theoretical usage" in which "degree of confirmation" measures the acceptability of scientific hypotheses. In Carnap's theory, a statement with a high initial probability may retain a still high probability in the face of an evidence statement that counts against it. But this is not to say that that the original statement is "confirmed by" the evidence statement.

Bar-Hillel conceded that Carnap's technical use of "degree of confirmation" is misleading, and that Popper was entitled to call attention to this "terminological oddity." He also admitted that, upon occasion, Carnap himself incorrectly equated "confirmation" and "acceptability." In *The Logical Foundations of Probability* (1950), Carnap stated that to say that the probability$_1$ of *h* on *e* is high means that *e* gives a strong support to the assumption of *h*, that *h* is highly confirmed by *e*.[18]

But this was a slip on Carnap's part. In its systematic usage in *The Logical Foundations of Probability*, "degree of confirmation" is not taken as a measure of the acceptability of hypotheses.

Popper Returns to the Attack

In a reply to Bar-Hillel, Popper insisted that his criticisms of Carnap's confirmation theory cannot be dismissed as a mere quibble over how to use the word "confirmation." He maintained that his objections to Carnap's theory are substantive.[19]

One such objection is the counterintuitive value assigned to universal hypotheses in Carnap's theory. In the case of universal laws for which infinitely many substitution instances are possible, c (H, e) = 0. Moreover, the value of the confirmation-function for a universe with a very large, but finite, number of substitution instances is indistinguishable from zero. On Carnap's theory, we are forced to assign c = 0 both to generalizations like "All ice cubes float in water," for which there is a great deal of evidence, and "All lead cubes float in water," for which there is no positive evidence. Carnap's theory thus leads to counterintuitive results for an important class of statements.

In addition, Popper argued that Carnap's theory is internally inconsistent. If correct, Popper's argument would be decisive against Carnap's theory of confirmation. Popper's argument is that there is a deductive consequence of the axioms of the probability calculus that is equivalent to the "special consequence condition," a condition that Carnap had shown to be false within his theory of confirmation. Popper noted that (1) below is a validly derived formula within the probability calculus:

(1) $P [(a \& b), z] \leq P (b, z)$.

If (a & b) = y and b = x, so that *y* entails *x*. then for every *z*:

(2) $P (y, z) \leq P (x, z)$.

But (2) has the same form as the "special consequence condition" that Carnap had proved false within his theory of confirmation, viz.,

(3) $C (y, z) \leq C (x, z)$,

which holds for all *z* if *y* entails *x*. (If *y* entails *x*, then every *z* that confirms *y* confirms *x* at least to the same degree as it confirms *y*.)

Since Carnap takes the relative logical probability of y with respect to z as a measure of the "degree of confirmation" of y, given z, his theory contains a contradiction.[20]

Carnap acknowledged that (2) is a theorem of the calculus of probability. But he maintained that Popper was wrong to hold that (3) is the "special consequence condition" proved false in *The Logical Foundations of Probability*. The special consequence condition is

(4) If y entails x, then, for all z, if C (y, z) > C (y, t), then C (x, z) > C (x, t), where t is a tautology.

Since Popper's argument depends on mistaking (4) for (3), it does not succeed.[21]

However, although Popper failed to demonstrate that Carnap's theory is internally inconsistent, he had raised the question of the appropriateness of a logical reconstructionist approach to the problem of confirmation. The construction of formal systems is a legitimate enterprise in logic and mathematics. It remains to be shown that these formal systems are applicable to the relation between scientific hypotheses and statements reporting evidence.

Lakatos' Appraisal of the Carnapian Program

In 1967, Imre Lakatos surveyed the history of the Carnapian program for inductive logic and concluded that it had been a "degenerating research program."[22] According to Lakatos, the Carnapian program passed through three stages:

1. a search for a logic of justification, in which the "degree of confirmation" of hypotheses is expressed in terms of the concept of "logical probability" (1930s);
2. a shift of focus from "degree of confirmation" to "qualified instance confirmation" (1940s); and
3. a shift of focus from "qualified instance confirmation" to "rational betting quotients" (from 1950).

On Lakatos' reconstruction, the transition from stage (1) to stage (2) is a response to the recognition that the degree of confirmation of universal statements is zero. Lakatos suggested that Carnap had four options at this point:

1. abandon the assumption that degree of confirmation is a logical probability;

2. define a different probability function—c*—such that C* (H, e) ≠ 0 for universal hypotheses;
3. show that it is not really counterintuitive that C (H, e) = 0 for universal statements; or
4. restrict the applicability of probability values to singular statements.[23]

By selecting option (3), Carnap directed his research program into stage (2). It is "qualified instance confirmation" that is important. If we reeducate our intuitions, we come to realize that what counts is whether the next new instance not included in *e* also is an instance of the hypothesis. This "qualified instance confirmation" of a universal hypothesis may approach 1.0 under suitable conditions. For instance, the qualified instance confirmation that the next raven observed will be black is very close to 1.0.

On Lakatos' reconstruction, the transition from stage (2) to stage (3) was undertaken to underline the distinction between "confirmation" and "reliability." Consider the hypothesis "All tossed coins show heads." This is an unrestrictedly universal statement, and its degree of confirmation is zero. But its "qualified instance confirmation" (or "reliability") is 1/2. A rational man, required to bet on the next toss of the coin, would assign a value 1/2 to the prediction that the next toss will show a head. In stage (3), Carnap exercised options (4) and (3) in the computation of "rational betting quotients."

Lakatos noted that, in order to compute "rational betting quotients," the inductive judge first must select a language within which evidence statements and hypotheses are to be expressed. How should the inductive judge proceed? The conservative position would be to select the language of the currently successful theory in the domain under consideration. But suppose a rival theory is proposed which reinterprets some of the terms of the entrenched theory. There is no decision procedure within the Carnapian program for the selection of one of two competing theoretical languages. Lakatos declared that

> the inductive logician can, at best, say to the scientist: "if *you* choose to accept the language *L*, then I can inform you that, in *L*, c (h, e) = q."[24]

For example, the inductive judge may select a language in which "mass" is a function of "velocity" (relativistic mechanics) or a language in which "mass" is not a function of "velocity" (classical mechanics). The estimation of "rational betting quotients" can be made only after selection of a language.

Lakatos insisted that the successive introduction of high-level theories is essential to scientific progress. Such theories are reinterpretations of the evidential base. Because he took the evaluation of such theories to be a central concern of the philosophy of science, he pronounced a judgment of sterility on the Carnapian program. The initial brash attempt to evaluate scientific hypotheses had been abandoned. In its place, Carnap had produced "a calculus of coherent beliefs on whose rationality he cannot pronounce."[25]

In response to this criticism by Lakatos, Alex Michalos suggested that, although the Carnapian program is inapplicable to cases of high-level theory-replacement, it is an appropriate reconstruction of "normal science" in which a dominant theory is deployed to cover additional phenomena.[26] According to Michalos, the principal evaluative task in periods of "normal science" is to establish the reliability of a hypothesis on the given evidence.

Lakatos maintained, however, that the Carnapian program is deficient even after a theoretical language has been selected. The estimation of "rational betting quotients" fails to do justice to striking new hypotheses that have minimal evidential support. Lakatos suggested that the Balmer formula for spectral lines in the hydrogen atom deserves higher marks after its application to just three spectral lines than Carnap's procedure would allow.[27] Lakatos also complained that Carnap's "rational betting quotients" fail to do justice to those instances that are particularly decisive tests of a theory.

In his 1954 essay, Popper maintained that the corroboration status of a theory should depend on the severity of the tests that it has passed. He emphasized that the severity, or ingenuity, of a test is not susceptible to precise quantitative measurement.[28]

Can Individual Hypotheses Be Justified?

In 1953, the Program Committee of the Eastern Division of the American Philosophical Association posed the following question about the justification of scientific hypotheses:

> is the process of justification one which inevitably brings us to justify total systems of knowledge rather than isolated statements?

The ensuing debate brought into sharp relief the logical reconstructionist orientation toward the philosophy of science.

Arthur W. Burks gave a negative answer to the Committee's question.[29] He conceded that the testing of a hypothesis often involves appeal to auxiliary hypotheses, and that a negative test result may be accommodated by modifying either the hypothesis under test or the auxiliary hypotheses. Burks insisted, however, that there is a second kind of testing of hypotheses—"probabilistic justification." He noted that in Carnap's theory, the logical probability of a hypothesis is independent of the status of other hypotheses in the language. According to Carnap,

$$c^* (h, e) = c_0 (h, e) / c_0 (e),$$

where "$c_0 (h, e)$" is the initial *a priori* probability (initial degree of confirmation) of *h*, given *e*, and c_0 is the initial *a priori* probability of *e*.

If Carnap's confirmation theory provides appropriate measures of the status of scientific hypotheses then a particular hypothesis can be justified without an appeal to other hypotheses. Burks maintained that Carnap's theory does provide appropriate measures of the confirmation-status of hypotheses. Of course, it does so by assigning initial *a priori* probabilities to the sentences of an artificial language.

Critics of Carnap's theory had denied that the concept of initial probability is relevant to the confirmation-status of scientific hypotheses. Against these critics, Burks argued that the very success of Carnap's theory establishes the relevance of the concept of initial probability. According to Burks, Carnap's theory is successful because it reproduces and renders more precise ordinary inductive arguments by simple enumeration and by analogy.[30] Since critics have failed to create alternative models that accomplish as much, the rational decision is to accept Carnap's confirmation theory as the best available reconstruction of the justification of scientific hypotheses.

Frederick L. Will complained that Burks had not established the relevance of formal models to the evaluation of scientific hypotheses.[31]

Indeed Burks' argument for the logical reconstructionist approach begs the question. Carnap's theory reproduces induction by simple enumeration because it is set up to do so. However, it is precisely the applicability of this form of inference to the justification of hypotheses that is in question.

Nor is it to the point for Burks to demand that dissenters propose alternative logical models. The central issue is whether the justification of hypotheses is to be understood in terms of formal models that specify probability-relations among sentences.

Will pointed out, in addition, that estimates of the logical probability of hypotheses are based on the assumed truth of evidence statements. Burks did not indicate how the truth of evidence statements is to be established. It is far from clear that this can be done without tacit appeal to theories. And if theoretical considerations are involved then it is not the case that a hypothesis can be justified independently of other hypotheses.

Will correctly emphasized that many scientific observation reports are instrument-dependent. The truth of such reports can be established only by appeal to theories about the functioning of instruments.

However, it yet may be the case that some observation reports can be established independently of theory. Will's opinion was that this is most unlikely. He declared that

> there is, in the use of a physical object statement, a reference, however implicit, to a complex kind of order, and ... the establishment of such a statement on the basis of certain selected features discriminated in observation consequently involves, whether one recognizes it, or not, a dependence upon that order.[32]

For example, to establish as true the statement that what was observed a moment ago was a ball rolling down an inclined plane requires reference to a set of assumptions about the behavior of such objects. These beliefs concern the properties a ball exhibits if rolled, if tested with calipers, if used to cast shadows, etc.[33] In speaking about "natural kinds"—different types of physical objects—we commit ourselves to certain presumptions about regularity.

Whether these presumptions should be called "theories" is questionable. The important point with regard to Burks' thesis is that physical-object statements can be established as true only upon appeal to presumed regularities.

In Carnap's confirmation theory, a hypothesis receives a degree of confirmation with respect to an evidence statement in isolation from other statements in the artificial language. In science, one can establish that an observation statement is true, and thus potentially confirming evidence for a hypothesis, only by reference to the complex relations present within a natural language.

Doubts about the Orthodox View of Theories

A Non-Statement View of Theories

The orthodox position was that a theory is a collection of statements. The axiom system is a deductively organized set of sentences, and the rules of correspondence also are sentences. Frederick Suppe, among others, opposed this position, suggesting that it be replaced by a "non-statement view" of theories.[1]

On the non-statement view, a theory is a non-linguistic entity that is expressed by alternative linguistic formulations. In this respect a theory is rather like a proposition. Many philosophers hold that two or more sentences may express a single proposition.[2] Examples include "It is raining," "Il pleut," and "Es regnet"; "it is false that Peter smokes" and "Peter does not smoke"; and, "John loves Marsha" and "Marsha is loved by John."

Proponents of the non-statement view claimed support from the work of John von Neumann. Von Neumann proved that Heisenberg's matrix mechanics and Schrodinger's wave mechanics are equivalent. It seemed plausible to conclude that quantum theory is a non-linguistic entity which is expressed by both the sentences of matrix mechanics and the sentences of wave mechanics.[3]

Suppe suggested that the linguistic formulations of a theory stipulate properties and relations of an idealized, perfectly isolated physical system whose states unfold according to relations specified by the interpreted theory.[4] Phenomena that fall under the intended scope of the theory are correlated with the idealized system thus characterized. Basic to this correlation are contrary-to-fact claims of the form "if the phenomena were accurately and fully characterized by the temporal unfolding of the idealized system, then such-and-such relations would be observed."

Theories and Explanation

It was the orthodox position that theories explain experimental laws by exhibiting the laws as deductive consequences of axioms and rules of correspondence. For example, Archimedes' theory explains the law that "ice floats in water" because that law is a deductive consequence of Archimedes' principle of buoyancy and statements about the physical properties of ice and water. In similar fashion the kinetic theory of gases explains Boyle's Law by exhibiting it as a deductive consequence of the axioms and rules of correspondence of the theory.

Ernest Nagel declared in *The Structure of Science* (1961) that

all explanations of laws seem to be of the deductive type.[5]

Nagel thus supported Pierre Duhem's thesis that theories explain laws by binding them together within a deductive system. Duhem had insisted that theories explain because they imply laws, and not because they depict some "reality" that underlies phenomena.[6]

However, it would appear that microtheories like the kinetic theory of gases explain macroscopic phenomena by reference to an underlying structure. Wilfrid Sellars complained that the orthodox position mistakenly reduces explanation to implication. Sellars maintained that what a microtheory explains is why observable things obey particular experimental laws to the extent that they do.

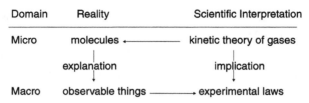

Figure 28 Sellars' Pattern of Microtheoretical Explanation.

For example, the kinetic theory explains why gas samples obey the law "P V / T = k." Sellars declared that

> roughly, it is because a gas "is"—in some sense of "is"—a cloud of molecules which are behaving in theoretically defined ways that it obeys the Boyle-Charles law.[7]

The actual behavior of a gas diverges from the ideal gas law at high pressures. This divergence also can be explained by appeal to the theory. The theory is "about" an "ideal gas," a gas comprised of point-masses that exert forces on one another only upon impact. No actual gas conforms to this idealization. As the pressure of a gas increases, its behavior diverges increasingly from the ideal behavior "described" by the theory.

Sellars thus challenged the deductive-subsumption view of the explanation of laws. He maintained that the basic schema of microtheoretical explanation is that

> molar objects of such and such kinds obey (approximately) such and such inductive generalizations because they *are* configurations of such and such theoretical entities.[8]

On this view, the search for an explanation of experimental laws is a search for an underlying mechanism at a deeper level of organization.

Models, Structural Isomorphism, and Substantive Analogies

Braithwaite had given a formalist interpretation of models. A theory and its model share a common structure. The deductive system and the model are isomorphic interpretations of the same calculus.

Braithwaite had conceded that a model may be of use in suggesting ways in which a theory may be augmented or modified. He also had maintained that theories explain by serving as premises for the deduction of laws. Since laws can be deduced from the axioms and rules of correspondence alone, the associated model is superfluous for the purpose of explanation.

In a paper published in 1958, Marshall Spector criticized the formalist interpretation of models. Spector complained that, on Braithwaite's view, both formal analogies and substantive analogies count as "models."[9]

An example of a formal analogy is the isomorphism between noneuclidean geometries and algebra. The axioms of hyperbolic geometry, for instance, can be mapped onto the axioms of algebra. The algebraic objects of the model and the geometrical entities of the theory participate in relations of the same form. Algebra thus is a "model" on Braithwaite's view. However, no other-than-formal relations hold between the algebraic objects and the geometrical entities.

Nor is a more-than-formal analogy present between electric-circuit theory and acoustical theory. Nevertheless, certain relations in electric current theory are formal analogues of relations in acoustical theory. Indeed, problems in acoustical theory sometimes can be solved by setting up an appropriate electric circuit and calculating acoustical parameters from their electrical circuit analogues.

The relationship of model and theory is different in the kinetic theory of gases. There is in the kinetic theory, as in the "algebraic theory of geometry" and the "electrical theory of acoustics," a structural isomorphism between "model" and theory. The model-objects and theoretical entities participate in relations of the same form.

In addition, the kinetic-theory model is a "candidate for reality."[10]

In the deductive system, certain defined terms of the calculus are associated with the pressure, volume, and temperature of a gas in its container. In the model, certain primitive terms represent the properties and motions of a collection of elastic spheres. For example p_a, a term of the calculus, is correlated with "pressure," an operationally defined term; p_a also is correlated, *via* the model, with the rate at which momentum is transferred to a wall of the container upon impact of the spheres. In classical dynamics, rate of momentum transfer per unit area is force per unit area, and force per unit area is pressure. Spector declared that

> thus we not only have a formal identity—a shared calculus—but also a *substantive* identity of two properties, one from the domain of the theory and one from the domain of the model.[11]

A natural question at this point is whether a similar substantive identity holds between the temperature of the gas and the mean kinetic energy of the elastic spheres. If it does, then it may be possible to measure the velocities of the elastic spheres directly. Experiments to accomplish this have been performed. The distribution of velocities in a beam of molecules emerging from a slit in an oven has been measured by a number of techniques.[12] The results correspond closely to the Maxwell-Boltzmann velocity distribution, which is a theorem of the deductive system.

The "billiard-ball model" for kinetic theory is not just another isomorphic interpretation of the calculus. Not only do gases behave as if they were composed of rapidly moving tiny elastic particles, they *are* so composed. Indeed, it was a consideration of the properties of the model that suggested the molecular beam experiment. This experiment was a *new* test of the kinetic theory of gases. Spector maintained that this is a case in which the theoretical entities of a deductive system have been identified with the corresponding objects of the model.[13]

One important aspect of this identification is a direct interpretation of the theoretical terms of the deductive system. The velocity distribution derived within the deductive system is identified with the velocity distribution of the model-objects. The pressure of the gas is identified with the pressure exerted by impacts of model-objects on the walls of the container. On the formalist interpretation, theoretical terms acquire empirical significance only "from below" *via* the upward-moving zipper. Spector concluded that the formalist position on the empirical meaning of theoretical terms is incorrect.

Hempel Abandons the Orthodox Conception of a Scientific Theory

The orthodox conception of a scientific theory is that it is an interpreted calculus. In itself the calculus is an uninterpreted set of sentences. The non-logical terms of the calculus are implicitly defined by its axioms. These theoretical terms acquire empirical meaning from below *via* the associated rules of correspondence. The rules of correspondence stipulate relations between certain of the theoretical terms and terms that refer to observables. Noretta Koertge labeled this orthodox thesis the "capillary model" of the meaning of theoretical terms.[14]

In an essay published in 1970, Carl Hempel conceded that the orthodox conception of a scientific theory—a conception that he previously had accepted— is incorrect.[15] It is incorrect because theoretical terms may acquire meanings in ways other than by "capillary rise" from the soil of the observation language. Hempel noted that a new scientific theory nearly always includes among its "theoretical terms" concepts that have a previous usage in other theories. Bohr, for instance, formulated a theory of the hydrogen atom in terms of such borrowed concepts as "nucleus," "orbit," and "energy." If a theory makes use of borrowed concepts in this manner, then it is inappropriate to regard the theory as an *uninterpreted* axiom system that subsequently receives an empirical interpretation. Hempel declared, moreover, that

> when at some stage in the development of a scientific discipline a new theory is proposed, offering a changed perspective on the subject matter under study, it seems highly plausible that new concepts will be needed for this purpose, concepts not fully characterizable by means of those antecedently available.[16]

For these reasons, Hempel now maintained that the problem of specifying how theoretical terms acquire empirical meaning is a "spurious problem."[17] It is a spurious problem because it is based on a mistaken presupposition. This presupposition is that one must begin with an uninterpreted calculus and then relate its terms to the concepts of an observational language. Hempel now rejected this presupposition.

Hempel's post-orthodox position was that a scientific theory is an ordered couple of two sets of sentences—"internal principles" and "bridge principles."—viz.:

$$T = < I, B >.$$

Internal principles specify the laws which govern the entities and processes posited by the theory. Bridge principles link this "theoretical scenario" to the phenomena to be explained.

In Bohr's theory of the hydrogen atom, for example, the internal principles specify permissible transitions between electron orbits in the atom. The bridge principles link these intra-atomic transitions to such terms as "hydrogen gas," "wavelength," and "velocity of light." These terms are part of the "antecedent vocabulary" to which the new theory is anchored.

Hempel's post-orthodox conception of a scientific theory—like the orthodox conception—presupposes the existence of two vocabularies. In the post-orthodox conception there is a theoretical vocabulary introduced by the internal principles and an antecedent vocabulary utilized in the bridge principles. Hempel emphasized that the theoretical vocabulary–antecedent vocabulary distinction does not correspond to the orthodox distinction between a theoretical vocabulary and an observational vocabulary. He declared that

> the terms of the antecedent vocabulary are by no means assumed to be "observational" in the sense of the familiar theoretical-observational distinction: that is, they are not required to stand for entities or characteristics whose presence can be ascertained by direct observation, unaided by instruments and

theoretical inference. Indeed, the antecedent vocabulary of a scientific theory will often contain terms which were originally introduced in the context of an earlier theory, and which are not observational in the sense just mentioned.[18]

Doubts about the Chinese-Box View of Scientific Progress

Feyerabend's Incommensurability Thesis

The Chinese-Box view of scientific progress was anathema to Paul Feyerabend. Feyerabend believed this view to be a caricature of science. Science is not a progressive incorporation of theories into more inclusive theories. Indeed, high-level theories are incompatible with their predecessors.

Feyerabend pointed out that the Chinese-Box view of scientific progress implies a "consistency condition," according to which

> only such theories are admissible (for explanation and prediction) in a given domain which either *contain* the theories already used in this domain or are at least *consistent* with them.[1]

The consistency condition, in turn, follows from two basic premises of logical reconstructionist philosophy of science:

1. A theory is acceptable only if it is confirmed to some degree by the available evidence.
2. If observation report O confirms both T and T^*, then T and T^* are logically compatible.[2]

Feyerabend argued that the examples of "reduction" cited by orthodox theorists fail to satisfy their own consistency condition. The transition from Galileo's physics to Newton's physics, for example, fails to satisfy Nagel's criterion of derivability. Galileo's law of falling bodies is not a deductive consequence of Newtonian mechanics. In Newtonian mechanics, gravitational force, and hence the mutual acceleration of two bodies, increases with decreasing distance. A body in free-fall toward the earth has a motion in which acceleration is increasing. Galileo's Law states that the acceleration is constant. The Newtonian axioms imply Galileo's Law only in conjunction with the factually false claim that "distance of fall/radius of earth = 0."[3]

Nor does Newtonian mechanics reduce to general relativity theory. The transition from Newtonian mechanics to general relativity fails to satisfy Nagel's criterion of connectability. The term "length" is a property term in Newtonian mechanics. "Newtonian length" is independent of signal velocity, gravitational attraction, and the motion of the observer who measures it. "Relativistic length," by contrast, is a relational term that is dependent on signal velocity, gravitational fields, and the motion of the observer.[4] Feyerabend concluded that, since "Newtonian length" and "relativistic length" are concepts of different logical type, Newtonian mechanics is not reducible to general relativity theory. He also maintained that Newtonian mechanics is not reducible to quantum mechanics, and that classical thermodynamics is not reducible to statistical mechanics.[5]

Hilary Putnam responded to Feyerabend's challenge by suggesting that what is involved in reduction is only the deduction from the new theory of a suitable approximation of the old theory. For example, classical geometrical optics is not deducible from (and hence is not reducible to) electromagnetic field theory, but a suitable approximation of classical geometrical optics is so deducible. Putnam emphasized that all that should be expected from a theory of reduction is a set of criteria for the reduction of an approximate version of the replaced theory.[6]

Feyerabend replied that this is not enough. Given G (geometrical optics) and E (electromagnetic theory), Putnam had admitted that G and E are inconsistent, but had insisted that there exists a theory G^* that does follow from E. Feyerabend was unimpressed. He pointed out that of course all the deductive consequences of E follow from E. This is trivial. However, the original interest in reduction had been an interest in a relationship between actual scientific theories.[7]

If Feyerabend's analysis of historical examples of "reduction" is correct, then it appears that a given observation report can be compatible with each of two inconsistent theories. Feyerabend suggested two ways in which this can happen. One way is for T and T^* both to be confirmed by the available evidence within a certain domain, but to have divergent consequences outside the domain for which there is as yet no evidence. This was the relationship in the late 1950s between the Copenhagen theory of quantum mechanics and the theory of David Bohm and J. P. Vigier. Both theories yield predictions that are confirmed in the quantum domain. However, the two theories had divergent consequences in the yet to be explored high-energy–short-time-interval region of subatomic interactions.[8]

It also is possible for an observation report to be compatible with two or more incompatible theories in virtue of the margin of error associated with processes of measurement. If O_1 is derived from T, and O_2 is derived from T^*, and both O_1 and O_2 fall within the margin of experimental error involved in determining observation report O, then O is compatible with both T and T^*. This "indeterminateness of facts" affords great latitude for theory-construction.

Feyerabend maintained that it is methodologically imprudent to rest content with a theory just because it has proved to be compatible with certain data. Instead he recommended a methodological imperative of theory-proliferation:

> You can be a good empiricist only if you are prepared to work with many alternative theories rather than with a single point of view and "experience." This plurality of theories must not be regarded as a preliminary stage of knowledge which will be at some time in the future replaced by the One True Theory. Theoretical pluralism is assumed to be an essential feature of all knowledge that claims to be objective.[9]

According to Feyerabend, the proper approach to a domain in which one theory reigns supreme is to invent alternative theories. He noted that, even if an alternative theory fails to replace the established theory, it may contribute to progress by suggesting additional tests of the entrenched theory. Often a really pertinent test of a theory becomes available only after a competing theory has been formulated.[10]

Feyerabend also recommended the creation of theories that contradict "well-established facts." Facts often gain status on the basis of concealed assumptions, and the invention of rival theories may bring into question these assumptions. Feyerabend discussed Galileo's strategy in attacking the geostatic theory of the universe.[11] It was a "well-established fact" that stones dropped from a tower fall straight to the base of the tower. Galileo nevertheless affirmed the Copernican hypothesis that the earth rotates on its axis. But it would seem that the observed motion of falling bodies refutes the Copernican hypothesis. Feyerabend gave Galileo credit for uncovering a concealed assumption in the geostatic argument. The geostatic argument is:

1. If the "real motion" of a tower-released stone is straight down to its base, then the earth does not rotate on its axis.
2. The observed motion of a body is identical to its real motion (implicit assumption).
3. Tower-released stones are observed to fall straight down to the base.

∴ The earth does not rotate on its axis.

Galileo maintained that it is only when combined with premise (2) that the observed motion of falling bodies provides evidence to support the geostatic hypothesis. Galileo proposed to reject premise (2) He pointed out cases in which the "real motion" of a body is not as it appears. To a traveler in a uniformly moving carriage, for instance, it is buildings and trees that appear to be in motion. Galileo suggested that the tower experiment is an analogous situation. We observe the linear motion to the base of the tower, but

> that part of all this motion which is common to the rock, the tower, and ourselves remains insensible and as if it did not exist.[12]

Premise (2) is false, and the geostatic argument is unsound. Feyerabend emphasized that scientific progress was achieved by Galileo's stubborn defense of a hypothesis that seemed to contradict "well-established facts."

Feyerabend insisted that successful high-level theories do not incorporate their predecessors. The Chinese-Box view of scientific progress is incorrect. Indeed Feyerabend recommended the following "Incommensurability Thesis":

> what happens when transition is made from a restricted theory T′ to a wider theory T (which is capable of covering all the phenomena which have been covered by T′) is something much more radical than incorporation of the *unchanged* theory T′ into the wider context of T. What happens is rather a complete replacement of the ontology of T′ by the ontology of T, and a corresponding change in the meanings of all the descriptive terms of T′ (provided these terms are still employed).[13]

Although Feyerabend did not specify explicitly what counts as a "change of meaning," he did indicate that the meaning of a term is a function of the theory in which it occurs.[14] To his critics, it appeared that Feyerabend was committed to the position that any change in the structure of a theory is also a change in at least some the meanings of the terms of the theory.

Dudley Shapere complained that an alternative axiomatization ought not count as a change in the meanings of the terms of a theory.[15]

Peter Achinstein emphasized that it would be stretching the phrase "change of meaning" beyond all usefulness if every alteration of a theory were to count as a "change of meaning" of the terms involved. His best illustration is perhaps the alteration of the Bohr theory of the hydrogen atom to permit the electron to occupy elliptical orbits around the nucleus. Surely the inclusion of elliptical orbits among those orbits permitted the electron does not alter the meaning of "electron."[16]

Achinstein and Shapere professed to be unable to see how incommensurable theories can be compared. Achinstein pointed out that if theories T and T^* are incommensurable, if no term in T has the same meaning as a term in T^*, then T^* cannot contradict T. But it would appear that a theory may deny what is asserted by some other theory. For example, the Bohr theory, in which the angular momentum of the electron is quantized, appears to be a denial of the classical theory of the electrodynamics of moving charged bodies.[17]

In reply Feyerabend correctly stated that Achinstein had defended the following thesis:

> it must be possible for two theories employing many of the same terms to be incompatible ... And this presupposes that at least some of the common terms have the same meaning in both theories.[18]

Feyerabend sought to refute this thesis by citing examples of pairs of competing theories which do not share any element of meaning. But surely this is not enough. Achinstein's thesis is that it is possible for competing theories employing common terms to be incompatible. This thesis cannot be refuted by citing instances of competing but incommensurable theories.

However, in the course of his discussion of specific incommensurable theories, Feyerabend did sharpen the issue. He admitted that not every change in a theory produces a change of meanings. For instance, given T = classical celestial mechanics, and $T^* = T$ with a slightly changed value of the strength of the gravitation potential, the transition from T to T^* does not involve any change of meanings. Feyerabend noted that although T and T^* assign different force-values in a given application, the difference in values is not due to the action of different kinds of entities.[19] He contrasted this transition with a transition from T to T' where T' is general relativity theory. In the latter transition, the meaning of "spatial interval" does change, supposedly because the entities referred to differ in T and T'.

Consistent with this contrast, Feyerabend suggested the following criterion of "change of meaning":

Type of Change in Theory $T_1 \rightarrow T_2$	Change of Meaning of the Terms of T_1 That Occur in T_2
Alteration of the system of classes to which the concepts of T_1 refer	Yes
Alteration of the extensions of the classes, but not the way in which the classes are stipulated	No

Feyerabend insisted that before this criterion can be applied we need to decide how to "interpret" theories. His own choice was to decide questions about meaning-change by appeal to the principles of the respective theories and not by appeal to similarities or differences at the observational level.[20] He noted that this way of interpreting theories makes possible the analysis and correction (from above) of the ways we express the results of observation. By contrast, the orthodox interpretation protects the "observation language" from the possibility of theoretical reinterpretation. Feyerabend maintained that the scientific revolutions of the seventeenth and twentieth centuries reflect his interpretation and not that of orthodoxy.

However, even if Feyerabend's proposed interpretation is accepted, the criterion of meaning-change is useful only if unique and definite rules of classification can be specified for the "entities" referred to by theories. Shapere emphasized the difficulties that arise upon application of this criterion. The rules must be sufficiently definite to permit an unambiguous classification. If competing rules of classification are available, then it must be clear which rule is used implicitly by the theory in question. Shapere expressed doubt that this can be achieved in the case of high-level theories.

> Are mesons different "kinds of entities" from electrons and protons, or are they simply a different subclass of elementary particles? Are the light rays of classical mechanics and of general relativity ... different "kinds of entities" or not? Such questions can be answered either way, depending on the kind of information that is being requested ... for there are differences as well as similarities between electrons and mesons, as between light rays in classical mechanics and light rays in general relativity.[21]

An important residual problem for Feyerabend is how theories can be compared. He noted that low-level theories may be evaluated by comparing their consequences with "what is observed." This can be done because there exist background theories to provide a common interpretation for the observational consequences of low-level theories. But high-level theories cannot be compared in this way. No theory-neutral observation language is available at this level. Rather, each high-level theory specifies its own observation language.

Nevertheless, Feyerabend maintained that there are rational procedures for the evaluation of conflicting high-level theories. He suggested three procedures of theory-evaluation.

The first procedure for assessing competing theories is to formulate a still more general theory within which occur statements that can be used to test the competing theories.[22]

The second procedure is to compare the internal structures of the competing theories. Two theories may differ with respect to the length of derivation from postulates to observation statements. They also may differ with respect to the number and severity of approximations made in the course of the derivations. Feyerabend suggested that, other things being equal, a smaller length of derivation and a smaller number of approximations is preferable.[23] Why this should be, he did not say.

The third procedure is to compare the observational consequences of theories with "human experience as an actually existing process."[24] Feyerabend recommended a "pragmatic theory of observation," according to which

> a statement will be regarded as observational because of the *causal context* in which it is being uttered, and not because of what it means. According to this theory, "this is red" is an observation sentence, because a well-conditioned individual who is prompted in the appropriate manner in front of an object that has certain physical properties will respond without hesitation with "this is red"; and this response will occur independently of the *interpretation* he may connect with the statement.[25]

An observer is caused to respond in certain ways by the characteristics of the observational situation and his prior conditioning. Feyerabend declared that

> we can ... determine in a straightforward manner whether a certain movement of the human organism is correlated with an external event and can therefore be regarded as an indicator of this event.[26]

Feyerabend proposed to use these verbal "event-indicators" to evaluate high-level theories. A given "event-indicator" may be consistent with one theory and inconsistent with a second theory.

Shapere complained that although Feyerabend denied that there could exist a theory-independent observation language, he made use of theory-independent observations to evaluate theories.[27] Shapere denied that these theory-independent observations could perform the function that Feyerabend assigned to them. Even if observation sentences do issue from certain situations as conditioned responses of the observer, they are mere uninterpreted noises. They have no more linguistic content than a burp.[28] Before such a sentence can count as a test of a theory, it must be interpreted. Feyerabend had maintained, however, that observational findings are subject to reinterpretation at the hands of successful new theories. If this is so, then observation sentences cannot decide the issue between competing theories. In short, Feyerabend cannot have it both ways. Either an appeal to observation sentences can decide the issue between competing theories and observation sentences are not subject to reinterpretation by the theories compared, or an appeal to observation sentences cannot decide the issue between competing theories and observation sentences are subject to reinterpretation by the theories compared.

Of course, one can decide to reject a theory upon the occurrence within experience of a conflict between expectations derived from a theory and an observer's response to an observational situation. This is to decide not to interpret the response from the standpoint of the theory. Feyerabend can select this option only at the expense of introducing disharmony into his recommended methodology, since he repeatedly emphasizes the importance of theoretical reinterpretations of statements about what is observed.

In later essays, Feyerabend suggested that the evaluation of incommensurable theories may rest on aesthetic considerations. Matters of taste are involved. But matters of taste are beyond argument. Feyerabend compared the evaluation of high-level theories to the evaluation of poems.

> Poems, for example, can be compared in grammar, sound structure, imagery, rhythm, and can be evaluated on such a basis . . . Every poet who is worth his salt compares, improves, argues until he finds the correct formulation of what he wants to say. Would it not be marvelous if this free and entertaining process played a role in the sciences also?[29]

Growth by Incorporation or Revolutionary Overthrow?

William Whewell, in his *History of the Inductive Sciences* (1837), had compared the growth of a science to the confluence of tributaries to form a river.[30] Whewell's tributary–river analogy is the original statement of the growth-by-incorporation thesis. Bohr's application of the correspondence principle and Nagel's theory of reduction are subsequent versions of this thesis.

Post-war critics of this overview complained that the history of science is better characterized by competition and survival. Theories do not flow smoothly into one another. On the contrary, when one high-level theory succeeds another, it often does so as the victor of a life-or-death struggle. The vanquished theory in such conflicts is buried, and not subsumed, by its successor.

Stephen Toulmin drew attention to the repercussions within science produced by a change in "Ideals of Natural Order," Ideals of Natural Order

> mark off for us those happenings in the world around us which do require explanation by contrasting them with "the natural course of events"—i.e., those events which do not.[31]

In Aristotelian physics, for example, motion must be accounted for by specifying the agency responsible for it. The continuing motion of an airborne projectile is an anomalous motion from the standpoint of the Aristotelian ideal. In Newtonian physics, by contrast, uniform rectilinear motion is inertial motion. It is only deviations from inertial motion—accelerations or decelerations—for which one must account. Projectile motion is not an anomalous motion from the standpoint of the Newtonian ideal. Toulmin emphasized that the Newtonian ideal did not subsume the Aristotelian ideal. Quite the contrary, the Newtonian ideal replaced the Aristotelian ideal.

N. R. Hanson likened the transition from one high-level theory to another to a *gestalt*-shift.[32] Before the transition, phenomena are viewed in a certain way. After the transition, the viewpoint has changed. This happened, for instance, in the transition from the geostatic theory of Aristotle and Ptolemy to the heliostatic theory of Copernicus and Galileo. Hanson suggested that heliocentric theorists like Kepler viewed phenomena in a new light.

Hanson contrasted the hypothetical experience of Tycho Brahe and Kepler at dawn. Tycho, a geocentrist, sees the sun rise from beneath the stationary rim of the earth. Kepler, a heliocentrist, sees the earth rotating beneath a stationary sun. Hanson maintained that it is the *gestalt* sense of seeing—"seeing as" and not "seeing that"—that is important in such cases. Tycho and Kepler both "see that" an orange disk increases its distance from a line bounded by green and blue. Nevertheless, Kepler sees the dawn differently than does Tycho.[33]

Kuhn on Scientific Revolutions and Paradigm-Shifts

Thomas S. Kuhn proposed a rational reconstruction of scientific progress that accommodates emphases on both incorporation and revolutionary overthrow. He made a distinction between "normal science" and "revolutionary science." During periods of normal science there is growth by incorporation. A dominant "paradigm" is progressively refined and expanded to cover new applications.

In the first edition of *The Structure of Scientific Revolutions* (1962),[34] Kuhn appeared to use the term "paradigm" much as Toulmin had used the phrase "Ideal of Natural Order." In the course of applying a paradigm, anomalies arise. Results expected from applications of the paradigm are not realized. After an accumulation of anomalies, an alternative paradigm may be put forward. If the new paradigm is a serious competitor, then a period of conflict ensues. If the new paradigm triumphs, then a revolution has taken place. The resultant picture of the history of science may be represented as follows:

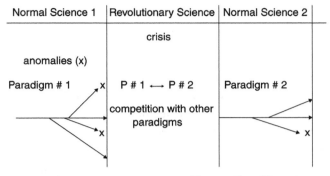

| Normal Science 1 | Revolutionary Science | Normal Science 2 |

crisis

anomalies (x)

Paradigm # 1   x | P # 1 ⟷ P # 2 | Paradigm # 2

competition with other paradigms

Figure 29 History of Science as a Succession of "Normal" and "Revolutionary" Stages. Source: John Losee, *Theories of Scientific Progress* (London: Routledge, 2004), 76.

Paradigm (1)	Anomalies	Paradigm (2)
Aristotle's physics	Free-fall; projectile motion	Newton's mechanics
Ptolemy's geocentrism	Phases of Venus	Galileo's heliocentrism
Phlogiston theory	Wt. calx >wt. metal	Oxygen theory
Wave theory of light	Partial reflection	Wave–particle dualism
Newtonian mechanics	Motion of Mercury	General relativity theory
Plenist models of the atom	α-particle scattering by gold foil	Nuclear model of the atom
Classical physics	Black-body radiation; photoelectric effect	Quantum physics

Feyerabend had maintained that the transition between high-level theories involves a wholesale replacement of one ontology by another. Every descriptive term of T_n that is still used in T_{n+1} has a changed meaning in T_{n+1}. Kuhn defended a similar position in the first edition of *The Structure of Scientific Revolutions*. He maintained that there is no paradigm-independent observation language, appeal to which decides the issue between competing paradigms.

Kuhn, like Hanson, emphasized the *gestalt* sense of seeing. In a famous passage, he declared that

> in a sense that I am unable to explicate further, the proponents of competing paradigms practice their trades in different worlds. One contains constrained bodies that fall slowly, the other pendulums that repeat their motions again and again. In one, solutions are compounds, in the other mixtures. One is embedded

in a flat, the other a curved, matrix of space. Practicing in different worlds, the two groups of scientists see different things when they look from the same point in the same direction.[35]

Israel Scheffler objected that Kuhn had reduced the history of science to a succession of viewpoints.[36] He pointed out that viewpoints may differ and yet refer to the same objects. Tycho and Kepler may "practice their trades in different worlds," but they both make referential claims about the changing positions of the same object—the sun.

Similarity of reference may provide a basis for paradigm selection. For example, the classical physicist and the relativity physicist make different predictions about the path of electrons in a synchrotron. Both predictions refer to the same objects, but the relativistic viewpoint clearly is superior. Hence, even though "electron" in special relativity theory—*qua* velocity-dependent mass—is a different concept than "electron" in classical physics, the two paradigms may be evaluated in virtue of referential equivalence. Kuhn had acknowledged that a gain in quantitative precision is a point in favor of paradigm replacement.

Gerd Buchdahl, Dudley Shapere, and others criticized Kuhn for vague and equivocal use of the term "paradigm."[37] In a "Postscript" added to the second edition of *The Structure of Scientific Revolutions* (1970),[38] Kuhn conceded that the criticism was justified. In the "Postscript" he distinguished two senses of "paradigm"—"paradigm" as "disciplinary matrix" and "paradigm" as "exemplar."

Kuhn held that a disciplinary matrix includes an

> entire constellation of beliefs, values, techniques and so on shared by members of a community.[39]

A disciplinary matrix includes commitments to types of inquiry, patterns and standards of explanation, inviolable principles, and the existence of theoretical entities. In addition, a disciplinary matrix typically includes a number of "exemplars." Exemplars—"paradigms" in the narrow sense—are particularly important successful applications of a theory.

Armed with this distinction between disciplinary matrices and exemplars, Kuhn surveyed the history of science. He concluded that

1. revolutions may occur within a micro-community of scientists without causing an upheaval within a scientific discipline; and
2. normal science may be pursued within a scientific community despite disputes over basic methodological and evaluative standards. Kuhn noted that chemists continued to conduct normal science throughout the

nineteenth century despite vigorous disagreement about the existence of atoms.[40]

By including these concessions in the "Postscript," Kuhn disarmed his critics. There no longer was present the hint of a history of science as a succession of observationally incommensurable viewpoints.

Kuhn's work promoted interest in the rational reconstruction of the history of science. Imre Lakatos and Larry Laudan made significant contributions to this project.[41] Preoccupation with this project diverted attention from the traditional problems of explanation and confirmation that had occupied philosophers of science working in the orthodox tradition.

The Legacy of Logical Reconstructionist Philosophy of Science

Logical reconstructionism dealt with methodological and evaluative problems by reference to logical relations among discrete levels of scientific language. Values of scientific concepts were related *via* operational definitions to statements of primary experimental data. These values were explained by formulating deductive (or inductive) arguments with laws as premises. Laws, in turn, were confirmed by logical relations to values of concepts, and ultimately to statements of primary experimental data. Theories explain laws by exhibiting them as logical consequences. Theory-replacement was taken to be progressive if the successor theory incorporates, and extends the range of, its predecessor. A goal of this analysis was an explication of epistemological concepts such as "explanation," "confirmation," and "justified theory-replacement."

The extensive and repeated criticisms of orthodox philosophy of science had a cumulative impact. Many philosophers of science came to believe that the disparity between actual scientific practice and the orthodox analyses of explanation, confirmation, and theory-replacement was too great. Perhaps logical reconstructionism is not the appropriate approach to methodological and evaluative problems in science.

Feyerabend versus Feigl on Logical Reconstructionism

Paul Feyerabend was eager to dismiss the logical reconstructionist program. He maintained that preoccupation with relations among the sentences of artificial languages is of no value to practicing scientists. He declared that there exists an enterprise which is taken seriously by everyone in the business where simplicity, confirmation, empirical content are discussed by considering statements of the form (x) (Ax ⊃ Bx) and their relation to statements of the form Aa, Ab, Aa & Ab, and so on, and *this* enterprise, I assert, has nothing to do with what goes on in the sciences.[1]

Feyerabend believed that he had shown that there is no theory-independent observation language. Scientists do compare theories and select the one they regard to be the most adequate. However, this is not a matter of measuring each theory against a theory-independent collection of observation reports.

On Feyerabend's view, orthodox philosophers of science have painted themselves into a corner. They have achieved irrelevance by becoming preoccupied with problems about "grue," counterfactuals and probabilities. Feyerabend insisted that the proper approach is to "return to the sources."[2] There is much to be learned about evaluative practice through the study of specific episodes from the history of science. Feyerabend applauded studies that had been completed by Thomas Kuhn, Vasco Ronchi, N. R. Hanson, and Imre Lakatos.[3]

Herbert Feigl also looked back on the career of logical reconstructionism. In opposition to Feyerabend, he concluded that there was much that had survived its demise. Consider the issue of theory-evaluation. Feigl admitted that orthodox theorists failed to provide a test-basis in a theory-independent observation language. This does not mean that "anything goes," as Feyerabend would have it.[4] There are thousands of low-level empirical laws that provide this test-basis. Feigl cited the laws of Balmer, Ampere, Coulomb, Faraday, and Kirchhoff, among others. In addition, there are thousands of laws governing chemical combination under specified conditions, and thousands of laws about the melting points, solubilities, specific heats, and electrical conductivities of various substances. These laws have passed numerous tests over decades and in some cases over centuries.[5] They are not incorrigible, but they have proved reliable in numerous applications.

Feigl conceded that the laws he had listed presuppose antecedently accepted conceptual frameworks. Moreover, many empirical laws presuppose theories about the operation of scientific instruments. Thus empirical laws are not formulated independently of all theoretical considerations.

Nevertheless, empirical laws often are free of infection from the high-level theories they support. Empirical laws are accepted independently of high-level theories, and these laws do provide support for selected theories. Illustrations from the history of science include:

Theory	Test-Basis
Mendelian genetics	Phenotype ratios
Kinetic molecular theory	Boyle's Law, Charles' Law ...
Bohr-Sommerfeld theory	Laws of Balmer, Brackett, Pfund ...
Optical isomerism	Laws of plane polarization
Plate-tectonic theory	Laws of geomagnetic induction

Empirical laws are not inviolable. They are not accepted independently of support from other propositions within science. Indeed, it is values recorded in observation reports that lead scientists to confer lawful status on certain correlations.

Individual empirical laws are subject to revision and even abandonment with the accumulation of experience. However, scientists accept the set of empirical laws at a particular time as a test-basis for theories.

Ernest Nagel had emphasized that whereas theories come and go, scientific laws have a life of their own.[6] Whatever theory of electromagnetism is accepted at a particular time, it must account for the laws of Ohm and Faraday. Theories account for experimental laws by exhibiting them as conclusions of deductive arguments. Feigl insisted that the deductive relationship between theories and laws is an emphasis of logical reconstructionism that survives its fall from grace.

Feigl argued, moreover, that the Chinese-Box view of the incorporation of one theory by another often is exemplified by transitions between theories in the history of science. An example is the transition from Galileo's theory of freely falling bodies, which Galileo applied as well to the motions of projectiles, to Newtonian mechanics. Newton's theory accounts for the successes achieved by Galileo's theory, and also accounts for many other empirical laws.

This is not a case of reduction, as Feyerabend had pointed out. Newton's theory corrects Galileo's theory "from above." It shows how Galileo's theory would be a deductive consequence in the limit as "distance of fall/radius of earth" approaches zero. Newton's theory not only incorporates Galileo's theory under limiting-case conditions, but also explains why Galileo's theory is not a logical consequence. Similar remarks apply to the incorporation of Newtonian mechanics into special relativity theory and the incorporation of classical thermodynamics into the kinetic theory of gases.

Feigl was unwilling to write off logical reconstructionist philosophy of science. He held that its achievements should engage the interest of practicing scientists.

Holton's Conclusion about Philosophy of Science

Perhaps these achievements *should* engage practicing scientists, but Gerald Holton observed that, as a matter of fact, contemporary scientists *do* ignore the recommendations of philosophers of science. According to Holton, the neglect of the philosophy of science by practicing scientists in the 1970s represents a shift in attitude toward the discipline. Earlier contributions to the discipline by Mach, Duhem, Einstein, Bohr, and Bridgman were highly regarded by practicing scientists. In part this was because these authors offered advice on evaluative problems from a platform of significant scientific achievements. More importantly perhaps,

> the immense forward thrust [of science] today is neither enlightened nor diverted by epistemological debates of the kind that engaged so much energy and attention in the past, through the first half of this century.[7]

Interlude: A Classificatory Matrix for Philosophies of Science

Philosophy of science is a normative discipline. However, "normative" is an ambiguous term. There is general agreement that philosophy of science is "normative$_1$," insofar as it includes standards applicable to a multiplicity of instances. In addition, many philosophers of science take their discipline to be "normative" in a stronger sense as well. In this stronger sense, a "normative$_2$" philosophy of science issues recommendations about "proper" methodological and evaluative practice. Implementation of selected standards is held to be necessary for the creation of "good science." It is important to recognize, however, that philosophy of science can be both "normative$_1$" (*qua* concerned with standards) and not "normative$_2$" (*qua* prescriptive).

A preliminary classification of philosophies of science may be effected by posing the question "should philosophy of science have normative-prescriptive intent?" Most philosophers of science, including Aristotle, Whewell, Popper, and Hempel, have answered "yes."

On the prescriptivist view it is not enough, for instance, to reconstruct Harvey's evaluation of Galen's theory of the blood, Whewell's evaluation of the theory of organic evolution, and Mach's evaluation of atomic theory. The philosopher of science must ask not only "what evaluative procedures have been practiced?" but also "what evaluative procedures ought be utilized to ensure scientific progress?"

Prescriptive philosophies of science may be located (roughly at least) on a line whose end-points are unqualified historicism and unqualified logicism. The unqualified historicist position is that historical considerations provide the sole warrant for evaluative standards. The unqualified logicist position is that logical and epistemological considerations provide the sole warrant for evaluative standards.

Of course the unqualified positions are unrealized abstractions. Actual philosophies of science appeal to both logic and history. Nevertheless, there are differences of emphasis. Consider the contrasting positions of Whewell and Mill. Whewell declared that

> the evidence in favour of our induction is of a much higher and more forcible character when it enables us to explain and determine cases of a kind different from those which were contemplated in the formation of our hypothesis.[1]

Whewell provided a historicist justification of undesigned scope. He maintained that there is no instance within the history of science in which a theory unexpectedly accounted for a class of facts (without specific adjustment for that purpose), and subsequently was shown to be false.

Mill, by contrast, argued that evaluative standards are justified by appeal to logic and epistemology. According to Mill, genuine scientific laws state connections that are both invariable and unconditional. Mill maintained that scientific laws may be distinguished from mere regularities by reference to a deductive pattern. Given law-candidates $L_1, L_2, \ldots L_n$ and empirical correlation p, L_1 achieves lawful status provided that the argument below is sound:

1. L_1 explains p and p is true.
2. $L_1, L_2, \ldots L_n$ are the only possible p-explainers
3. Neither L_2 nor L_3 nor $\ldots L_n$ explains p.
4. If [(1) and (2) and (3)] then L_1 is true.

$\therefore$ L_1 is true.

Mill believed that there are cases of successful application of this pattern, despite the obvious difficulty of establishing the truth of the premises. But although Mill claimed that the pattern often is instantiated successfully, the only example he cited was Newton's law of gravitational attraction.[2] Newton's $1/R^2$ force law explains the orbits of the planets (given rectilinear inertial motion). Force laws with other values of exponential dependence do not.

Prescriptivists and descriptivists disagree about the value of a commitment to inviolable principles. A principle is inviolable within a philosophy of science provided that those who practice that philosophy of science:

1. exclude the principle from the possibility of rejection or modification;
2. accept and apply the principle independently of historical considerations about the development of science; and
3. maintain that application of the principle is important in the appraisal of evaluative practice in science.

Examples of inviolable principles are:

1. Hempel's "equivalence condition." Hempel maintained that it is a necessary condition of the confirmation-relation that if evidence statement *e* confirms hypothesis *h* then *e* also confirms every *h** logically equivalent to *h*;[3]
2. Lakatos' criterion "incorporation-with-corroborated-excess-content."[4] Lakatos applied this criterion in the appraisal of both scientific research programs and historiographical research programs.
3. Laudan's methodological procedure (*Progress and Its Problems*, 1977) for the appraisal of competing philosophies of science.[5] The procedure features a determination of the "preferred intuitions" of the scientific elite about "standard cases" of scientific progress.

A philosophy of science either does or does not have normative-prescriptive intent, and either does or does not include principles held to be inviolable. Given these divisions, the possibilities for combination may be represented by the following matrix, in which *N* stands for normative-prescriptive intent and *I* stands for commitment to the existence of inviolable principles within philosophy of science.

	N	~N
I	N&I	~N&I
~I	N&~I	~N&~I

It would seem that the (~N&I) box must be empty. No purely descriptive philosophy of science can include trans-historical inviolable principles. By contrast, the (~N& ~I) and (N&I) boxes have numerous entries. The descriptivist positions of Holton and Hull fall within the (~N&~I) box. Foundationalist positions fall within the (N&I) box. The foundationalist maintains that there exist trans-historical, inviolable evaluative principles. These principles are foundational in the sense that (1) they are accepted independently of other statements in science, and (2) the evaluative status of these other statements depends on the foundational subset. Logical reconstructionism is a resolutely foundationalist position. Logical reconstructionists proposed trans-historical, inviolable standards for evaluative tasks such as confirmation, explanation, and theory-replacement. The controversial box is (N&~I). Some proponents of normative naturalism affirm prescriptive intent but deny that any evaluative principle is inviolable. An important question is whether there can be a normative-prescriptive philosophy of science devoid of inviolable principles. If not, then the various versions of naturalism actually fall within either the (N&I) box or the (~N&~I) box.

Descriptivism

One response to the extensive criticisms of logical reconstructionism was a reconsideration of the nature of the philosophy of science. Some philosophers of science maintained that the proper aim of the discipline is the description of scientific evaluative practice. Descriptive philosophies of science fall within the (~N & ~I) box.

There is a modest version and a robust version of descriptive philosophy of science. The aim of the modest version is the historical reconstruction of actual evaluative practice. Given that scientists preferred one theory (explanation, research strategy . . .) to a second the modest descriptivist seeks to uncover the evaluative standards whose application led to this preference. For instance, the modest descriptivist may seek to uncover the standards implicit within such evaluative decisions as Aristotle's rejection of pangenesis, Newton's rejection of Cartesian vortex theory, or Einstein's insistence that the Copenhagen interpretation of quantum mechanics is incomplete. Pursuit of a modest descriptive philosophy of science may require a certain amount of detective work, particularly for episodes in which the pronouncements of scientists and their actual practice do not coincide.

The conclusions reached by modest descriptive philosophy of science are subject to appraisal by reference to standards applicable to historical reconstruction in general. There is no distinctively philosophical task of appraisal. The modest descriptivist is a historian with a particular interest in evaluative practice.

The robust version of descriptive philosophy of science derives from, or superimposes upon, the conclusions of modest descriptivism, a *theory* about evaluative practice. The theory is put forward as a contribution to our understanding of science. It purports to explain why science is as it is. A robust descriptive philosophy of science typically includes the claim that scientific evaluative practices exhibit certain patterns or conform to certain principles. Of course, not every historical instance will exhibit a pattern exactly or conform precisely to the requirements of a principle. But a successful robust descriptive theory must help us to understand at least some important episodes from the history of science.

Prescriptivists have argued that descriptivism ought be rejected because it fails to establish a distinction between (1) "evaluative practice proceeded thusly," and (2) "it did so proceed, but a different evaluation would have been superior." This argument begs the question against descriptivism. It presupposes what is to be proved, namely that descriptivists are mistaken about the appropriate aim of philosophy of science.

Descriptivists, in turn, have urged that prescriptivism be rejected because it is scientists, and not philosophers of science, who possess the relevant expertise in evaluative contexts. Gerald Holton provided support for this position by pointing out in 1984 that the leading scientists at that time, unlike scientists in decades past, take little interest in normative-prescriptive philosophy of science.[1] It is scientists themselves who are on the evaluative firing line, and they are no longer disposed to cede authority on these matters to second-order theorists. If Holton is correct about this change of attitude on the part of scientists, then this is a point in favor of the descriptivist approach to philosophy of science. Holton's argument does not beg the question against prescriptivism. How much weight it ought be accorded is another matter.

Meehl's "Actuarial" Approach

Paul Meehl has promoted an "actuarial" approach to philosophy of science. He observed that philosophers of science agree that there are several, sometimes conflicting, standards for the appraisal of scientific theories. Among these standards are simplicity, agreement with observations, breadth of scope and predictive novelty. Unfortunately, we have only sparse data on the relation between the fate of individual scientific theories and the presence or absence of these factors. Meehl recommended the accumulation of the appropriate data for randomly selected theories, and the subsequent appraisal of this data by the techniques of statistical analysis.

Meehl thus outlined a program for future research. He contrasted his "actuarial" program with the "subjective, impressionistic"[1] appraisals of those philosophers of science who draw conclusions about evaluative practice from selected case studies. Whether or not evaluative situations in science are amenable to the sort of statistical analysis Meehl proposes remains to be shown.

The actuarial approach assumes that theories can be individuated, and that the evaluative histories of theories T_m and T_n are equally important. Both assumptions are subject to challenge. But the recommendation to accumulate additional data on evaluative practice in science is sound.

Meehl took philosophy of science to be something more than the description and statistical analysis of actual practice. He subscribed to the robust version of descriptive philosophy of science. In its "robust" version, descriptive philosophy of science is a "science of science." As such, it is a discipline distinct from the history of science and the sociology of science. Meehl declared that the philosopher of science ought seek to *explain* the phenomena of the domain. The goal should be to understand why science "works," not merely to ascertain the empirical correlates of its working well or badly.[2] Meehl thus introduced the methodological equivalent of the relationship between phenomenological laws (e.g., Boyle's Law) and theories about underlying mechanisms (e.g., the kinetic theory of gases). Application of the actuarial program uncovers the phenomenological laws of evaluative practice. The philosopher of science then develops a theory about the underlying mechanism, thereby explaining "why science 'works.'" Meehl observed that since the "phenomenological laws" have not yet been uncovered, it would be premature to put forward a theory of science.

Holton on Thematic Principles

Gerald Holton has pursued the case study approach criticized by Meehl. However, he has done so not to find illustrations of the application of selected evaluative standards, but to describe evaluative practice within its historical setting. Holton has published studies of evaluative decisions made by Kepler, Newton, Mach, Bohr, Einstein, and Stephen Weinberg, among others.[1]

Holton found that the evaluative decisions of these scientists display commitments to various "thematic principles." Thematic principles include explanatory goals, directive principles, evaluative standards, ontological assumptions, and high-level substantive hypotheses.

Explanatory goals include the "Ionian Enchantment"—the ideal of a unified theory of all phenomena—and Niels Bohr's principle of complementarity. Directive principles include "seek quantities within phenomena that are conserved, maximized, or minimized," and "seek to interpret macroscopic phenomena by reference to theories of micro-structure." Evaluative standards include simplicity, inclusiveness, and the display of undesigned scope. Ontological assumptions include atomism and plenism. And high-level substantive hypotheses include energy quantization, the discreteness of electric charge, and the constancy of the velocity of light.[2]

Thematic principles are a varied lot. They range from claims about the existence of certain kinds of entities to presuppositions about the aims of the discipline. Holton is surely right to emphasize their importance in scientific evaluative practice.

Holton did not issue prescriptive recommendations on behalf of selected thematic principles. His approach to the interpretation of science was resolutely descriptive. From this standpoint, he acknowledged that the corpus of thematic principles evolves within the history of science. Principles such as undesigned scope and complementarity are relatively recent additions to the corpus, the principle of conservation of mass has been modified, and the principle of conservation of parity has been abandoned.

Holton did place one prescriptive demand on the study of scientific evaluative practice. He insisted that the evaluative decisions of scientists be viewed within a three-dimensional interpretive framework. The axes of this framework represent empirical content, conceptual relations, and thematic content. Holton emphasized that a history of evaluative practice that omits the thematic dimension misrepresents the development of science.

It has been well documented that science is a cooperative and cumulative enterprise, that eminent scientists often disagree about the importance of particular research programs, and that successful theorists often "suspend disbelief" in the face of negative evidence. Holton maintained that these aspects of scientific enterprise are best understood by reference to thematic considerations.

In the first place, it is scientists' shared commitments to certain thematic principles that makes science a cooperative, cumulative undertaking. Scientists very often agree about the types of theory to be sought and the types of explanation that are relevant. This agreement promotes continuity within research.

It also is true, however, that commitments to thematic principles sometimes promote conflict. Certain thematic principles pull in opposite directions. For example, atomism and plenism are mutually exclusive ontological positions, and agreement-with-observations and simplicity are antithetical evaluative standards. It is commitments to opposed thematic principles that underlie the disputes between such gifted scientists as Newton and Leibniz, Bohr and Einstein, and Heisenberg and Schrodinger.

Holton observed that

> if science were two-dimensional, the work in a given field would be governed by a rigid uniform tradition or paradigm. But the easily documented existence of a pluralism at all times points to the fatal flaw in the two-dimensional model.[3]

A third feature of the history of science that remains mysterious on a two-dimensional model is the propensity of scientists to perform a "suspension of disbelief." Holton called attention to Einstein's response to Walter Kaufmann's experimental result (1906) that appeared to falsify the special theory of relativity. Einstein refused to accept the finding as decisive. According to Holton, he did so largely because he was strongly committed to the thematic principles of simplicity, unification, symmetry, and invariance.[4] Had Einstein not been convinced that fulfilling these formal requirements is of overriding importance, he presumably would have attributed greater weight to the *prima facie* destructive implications of Kaufmann's work. A decade later Kaufmann's result was shown to be incorrect.

Holton's three-dimensional model is a framework for the interpretation of the history of science. It incorporates, and improves upon, William Whewell's fundamental interpretive polarity of fact and idea. Holton maintained that application of the three-dimensional model enables us to understand why science displays an overall pattern of continuity, punctuated by occasional methodological disputes and instances of "suspension of disbelief." By utilizing the three-dimensional model as a premise in explanatory arguments, Holton has formulated a robust descriptive philosophy of science.

Holton developed a theory of scientific evaluative practice without relying on comparisons with other processes or activities. Other robust descriptive philosophies of science have been based upon analogies to other processes or activities. For instance, Ron Giere has formulated a decision-theoretical model of the evaluative activities of scientists, Paul Thagard has developed an analogy between theory assessment and the operation of computational programs, and David Hull has suggested that certain aspects of scientific evaluative practice are the effects of a "natural selection."

Giere's "Satisficing" Model

Giere recommended a two-stage program for descriptive philosophy of science: (1) discover how scientists do in fact make evaluative decisions, and (2) provide a theoretical interpretation of the decisions rendered. He suggested a "satisficing" variant of decision theory to explain theory-choice in science. He insisted that the appropriate version of decision theory is "descriptive" rather than "rational." A "rational" decision theory seeks to formulate a uniquely rational strategy for decision making. A "descriptive" decision theory, by contrast, uncovers the decision strategies that actually are implemented by scientists. It seeks to "translate" the beliefs and desires of agents into utility measures applicable to outcomes of alternative evaluative decisions.

On the "satisficing" model, the agent (1) sets a minimum satisfactory payoff, (2) surveys the range of available decision options, (3) selects an option that fulfills the minimum payoff condition (if any exist), and (4) acts accordingly. Giere noted that if there is no option leading to a satisfactory payoff, then the agent either lowers the satisfaction level or changes some aspect of the decision situation.

Giere suggested that the geologists who opted for continental-drift theory in the 1950s were acting as "satisficers." He conceded, however, that presentation of a single example is insufficient to create widespread acceptance of the decision-theoretical approach to the philosophy of science. He expressed hope, however, that the example would encourage other analysts to pursue a similar approach.[1]

Thagard's Computational Philosophy of Science

Paul Thagard recommended a program for a "computational" account of scientific evaluative practice.[1] The acknowledged goal of the program is the formulation of normative-prescriptive evaluative standards. Thagard suggested, however, that a descriptive philosophy of science be developed first and that normative conclusions be drawn subsequently.

In its descriptive (first) phase, computational philosophy of science seeks answers to questions about the process of scientific discovery and the appraisal of scientific theories and explanations. The aim is to achieve an understanding of science by developing analogical links to the operation of computational processes. This presupposes, of course, that science is an activity that stands in need of an explanation.

The basic resemblance-relations for a computational interpretation of mental processes associate ideas and data structures, and inferences and algorithms. Building upon this basic analogy, Thagard translated observation reports into "messages" and laws into "rules" of conditional form. He showed that a scientific theory may be represented as an interrelated set of concepts, rules, and problem-solutions. He also showed that upon translation of goals and background knowledge into a computational language, the formulation of new explanatory hypotheses can be simulated.

Thagard provided an illustration. He utilized computer program PI ("processes of induction") to answer questions about the nature of sound. Application of PI to initial data about propagation, reflection, and pass-through leads to the conclusion that sound is a wave motion.[2] Operation of the computational program thus simulates discovery of the wave theory of sound.

In a similar vein, Herbert Simon has shown that operation of computer program BACON upon data on planetary distances and periods leads to Kepler's Third Law, and that operation of computer program DALTON upon data that include combining volumes in gas-phase reactions yields appropriate molecular formulas.[3]

These achievements provide some support for the program of computational philosophy of science. However, Thagard conceded that much remains to be done if the computational approach is to succeed in reproducing patterns of scientific discovery. In particular, it will be necessary to provide more sophisticated techniques for the introduction of new concepts and the application of domain-specific heuristic principles.[4] The situation is even more complex for evaluative contexts in which multiple theories and explanatory arguments are in competition. Nevertheless, Thagard maintained that the computational program displays great promise.

Cognitive Development and Theory-Change

Alison Gopnik has drawn attention to similarities between the cognitive development of young children and theory-change in science.[1] She observed that three-year-olds group both physical objects and mental entities into "natural kinds." Among the mental entities are desires and beliefs. Children of this age regularly make predictions and give explanations about the behavior of other children.

In addition, three-year-olds apparently take perception to be veridical. Consider a case in which a group of three-year-olds are shown that a box contains pencils, and then are told that Betsy believes the box to contain chocolates. When asked to predict Betsy's behavior, they interpret Betsy to act as if the box contained pencils. Evidently there is no place in the three-year-olds' theory for behavior caused by false beliefs.

Gopnik noted that by age five the situation has changed. At that age children have replaced the earlier theory with a theory that interprets false beliefs to be efficacious in guiding behavior. She noted, in addition, that there occurs a transitional period in which older three-year-olds express interest in experimenting in puzzling cases.

Gopnik has drawn some far-reaching conclusions from studies of this type. She maintained that the cognitive development of young children is characterized by the following features:

1. children replace one theory with another;
2. in this process, they initially ignore counter-evidence to the original theory;
3. children subsequently account for counter-evidence by adopting auxiliary assumptions;
4. they next apply new theoretical ideas in limited contexts; and
5. they eventually reorganize their knowledge around the new theoretical concepts.

Gopnik claimed that cognitive development shares important similarities with scientific progress. She declared that

> the moral of my story is not that children are little scientists but that scientists are big children.[2]

Gopnik clearly intended the "scientist-as-big-child" theory to be a theory about science. By developing the analogy, one supposedly gains insight into the structure, functions, and dynamics of scientific theories.

In addition, Gopnik issued a causal claim. There are similarities between the cognitive development of young children and theory-change in science because science

> exploits powerful and flexible cognitive devices that were designed by evolution to facilitate learning in young children.[3]

Gopnik's program has drawn two types of criticism. The first type of criticism is that the analogy to cognitive development does not elucidate the dynamics of theory-change in science. Granted that children construct, revise, and replace "theories," this does not tell us anything about scientific theory-change that we did not already know. A robust descriptive philosophy of science is a theory about science and as such ought contribute to our understanding of science. Gopnik accepted this demand. She conceded that in order for the theory to be more than just a metaphor there has to be some interesting, substantive, cognitive characterization of science, independent of phenomenology and sociology.[4]

But the "interesting, substantive, cognitive characterization of science" has not yet been forthcoming.

Ron Giere pointed out a further difficulty. Babies of 1492 and babies today have similar cognitive capacities that are designed to "get many things right about the world in which we grow up." Five hundred years is a mere instant within evolutionary history. And yet the science practiced by adults in 1540 was very different from the science practiced today. Modem science features skeptical appraisal (and not mere pattern-matching), experimental methodology, and instrumental innovation. Science prior to 1540 did not. It would seem that our early childhood accommodation to environmental pressures tells us very little about scientific methodology.[5]

Gopnik replied that what it does tell us is that in the seventeenth century, technological and sociological advances allowed us to take inferential capacities that were designed to apply to certain problems in childhood and instead apply them to new problems in adulthood.[6]

But to reply in this way is to redirect our attention to the "technological and sociological advances" that made the difference between superstition and science.

The second type of criticism is that the cognitive development of young children is subject to other interpretations. Susan Carey and Elizabeth Spelke observed that cognitive development involves applications of schemata, lists, scripts, and prototypes "not usefully categorized as theories." They suggested, in addition, that much cognitive development can be accounted for by application, not of "theories," but of innate, unchanging "core conceptual systems."[7] These conceptual systems are response-specific and task-specific. They are involved, for example, in the perceptual discrimination of colors and depth, and the associated activities of scanning, grasping, and locomotion.

Carey and Spelke maintained that since we share these innate systems with other animals, we cannot account for the rise of science by acknowledging their role in cognitive development. They emphasized, moreover, that core conceptual systems, unlike theories, are not subject to replacement. They are elaborated during cognitive development, but they are not revised with the accumulation of experience. If Carey and Spelke are correct, there are too many important disanalogies between cognitive development and theory-change to sustain a "scientist-as-a-big-child" view.

The present author is not competent to adjudicate this dispute over the nature of cognitive development. However, I do believe that a caution is in order. Gopnik began the essay "The Scientist as Child" with a general characterization of the structure, functions, and dynamics of scientific theories, next called attention to putative examples of such theories in the cognitive development of young children, and then argued that the similarities between cognitive development and theory-change in science are significant. This sequence becomes circular if the investigator reads her antecedent understanding of scientific theories into her studies of the behavior of children.

Of course, the psychologist, like scientists in other fields, brings general assumptions and predispositions to her investigations. Francis Bacon's ideal of pure inquiry, conducted by scientists untainted by Idols of the Tribe, Cave, Market Place, or Theatre, is not an ideal that can be fully realized. But there is a further concern about Gopnik's program. Her interpretation of the behavior of children may be affected by quite specific commitments about the nature of scientific theories.

The Evolutionary-Analogy View

A number of philosophers of science have sought to account for the growth of science by reference to the theory of organic evolution. Michael Ruse has observed that "evolutionary philosophies of science" may be subdivided into two types:[1]

1. The "evolutionary-analogy" view develops an analogy between the growth of science and the operation of natural selection upon a pool of variants subject to environmental pressures. On this view, competing concepts, theories, and methods of inquiry engage in a competitive struggle from which the "best adapted" emerge victorious.

2. The "evolutionary-origins" view attributes the growth of science to the application of epigenetic rules that have proved adaptive within the course of evolutionary history. On this view, science develops as it does because certain methodological rules and evaluative principles have become encoded in our genes. Acting on these rules and principles presumably proved adaptive for our proto-human ancestors. The evolutionary-origins view—qua robust descriptivism—is a theory about the origins of modern science.

Toulmin on the Evolution of Scientific Concepts

Stephen Toulmin maintained that the development of scientific concepts is an evolutionary process in which the fittest conceptual variants survive. Just as ecological settings produce pressure on organisms to adapt, so also observation reports produce pressure on the concepts introduced to explain them.

Toulmin shared the interests of Kuhn, Lakatos, and Laudan in the rational reconstruction of the growth of science. He sought to superimpose the concepts of the theory of organic evolution upon the history of science. He observed that

> the central Darwinist insight was the recognition that . . . both the continuity of organic species, and their manner of change, can be explained in terms of a single dual process of variation and selective perpetuation.[2]

Toulmin suggested in *Foresight and Understanding* (1961) that in science, as in the evolution of biological species, "change results from the selective perpetuation of variants."[3] Subsequently, in *Human Understanding* (1972), Toulmin outlined a program for a systematic reconstruction of scientific progress in categories borrowed from the theory of organic evolution.[4] He maintained that important questions in science often take the form:

> given that concepts c_1, c_2, . . ., are in some respect *inadequate* to the explanatory needs of the discipline, how can we modify/extend/restrict/qualify them, so as to give us the means of asking more fruitful empirical or mathematical questions in this domain?[5]

Toulmin sought to shift attention from scientific theories to the concepts they contain. He declared that

> questions about scientific *concepts* underlie (are "logically prior to") questions about scientific *propositions*.[6]

This is true in the trivial sense that theories are about concepts. There are no concept-free theories. It does not follow from this fact alone that the history of a scientific discipline is represented more adequately as a struggle among competing concepts than as a struggle between competing theories.

Ernan McMullin opposed this shift of focus from theories to concepts. He insisted that

> the units of change in science are *not* concepts but concept-sets, i.e. theories. It is rarely the case that concepts alter in isolation, without effect on one another.[7]

Toulmin conceded that concepts are interrelated in various theories. Concepts may vary systematically together with other concepts. On the other hand, he insisted that concepts also are "free to vary independently"[8] despite the fact that they are interrelated within a specific theory at a particular time. Toulmin called for a shift of focus in the interpretation of science

> from "fuller knowledge through more true propositions" to "deeper understanding through more adequate concepts."[9]

In support of this thesis, he drew an analogy between organic evolution by natural selection and the growth of science by means of a selective competition among concepts.

Toulmin held that conceptual development within a scientific discipline is a "natural selection" that operates on a set of "conceptual variants." He noted that

> an evolutionary account of conceptual development . . . has two separate features to explain; on the one hand, the coherence and continuity by which we identify disciplines as distinct and, on the other hand, the profound long-term changes by which they are transformed.[10]

The rational reconstruction of scientific growth thus is an analysis of "intellectual ecology over time." As such, the philosophy of science is related to the history of science as ecology is related to phylogeny in evolutionary biology (see Figure 30).

Evolution of Species		Conceptual Change	
Phylogeny	**Ecology**	**History of Science**	**Philosophy of Science**
From what succession of precursors has this species descended?	By what sequence of responses to environmental pressures did the species acquire its present form?	From what succession of precursor concepts has this set of concepts descended?	By what sequence of responses to disciplinary pressures did this set of concepts arise?
A tree of descent.	Applications of the theory of natural selection.	A history of a scientific discipline.	A rational reconstruction of scientific growth.

Figure 30 Biological Change and Conceptual Change.

Source: Stephen Toulmin, *Human Understanding* (Oxford: Blackwell, 1986), 133–44.

instantiates the concept "the chemical properties of the elements are in periodic dependence on their atomic weights." By contrast, to interpret the sequence of proposed periodic arrangements as a lineage, in Hull's sense, seems a bit forced.

Hull's descriptive philosophy of science is robust. He maintained that we may increase our understanding of evaluative practice within science by reference to the general theory of selection processes. Indeed, he claimed that otherwise puzzling aspects of evaluative practice are readily understood when viewed from the standpoint of the general theory of selection processes.

One such aspect is the intensity of priority disputes among scientists. A scientist-interactor's *fitness* is measured by his or her contribution to successful lineages of scientific concepts. Any challenge to a scientist's claim to have contributed to a lineage is a serious threat indeed. It is not surprising that disputes over who first developed or presented an idea often generate heated controversies.

A second aspect of science accounted for on Hull's theory is the effective self-policing that prevails within the scientific community. Individual scientists are interactors who strive to have their ideas replicated within a selective environment. The mark of success (i.e., "fitness") is the publication, or other dissemination of results that are used subsequently by other scientists.

It is in the best interest of the individual scientist to play by the rules of the community. A scientist who falsifies data, uses the results of other scientists without proper attribution, or misinterprets the theories of other scientists undermines the process by which lineages of scientific concepts are created. Such behavior brings forth ostracism from the community. It greatly reduces the miscreant's likelihood of achieving fitness through replication in the future.

Scientists are fully aware of the consequences of misbehavior. Misbehavior therefore is uncommon. Contrast the effective disciplinary ethos of the scientific community with the failures of police organizations, medical organizations, and religious organizations to discipline their errant members.

Dawkins on "Genes" and "Memes"

Richard Dawkins has restated the theory of evolution in terms of the concepts "replicators" and "vehicles." Replicators are entities—typically genes—of which copies are made. Most copies are accurate. However, no copying process is perfect. Variation inevitably arises within a population of replicators. Those variants that prove best able to survive and make copies of themselves become dominant within the population.

Replicators thus are the fundamental units of natural selection. They are

the basic things that survive or fail to survive, that form lineages of identical
copies with occasional random mutations.[27]

Dawkins observed that the first replicators most likely were molecules. These
replicators were in competition with one another. At some point molecule-
replicators protected themselves by building "walls" around themselves.

This is where vehicles came into play. Vehicles are "large communal survival
machines"[28] created by the association of replicators. The bodies of plants and
animals are typical vehicles for the propagation of replicators. Natural selection
favors those replicators that create the most effective vehicles.

The gene is the replicator in sexually reproducing vehicles. Dawkins
characterized a gene as

any portion of chromosomal material that potentially lasts for enough
generations to serve as a unit of natural selection.[29]

Stephen Jay Gould objected to this shift of emphasis from individual
organisms to genes. He defended Darwin's thesis that natural selection acts on
individual organisms. According to Darwin, those individuals that are stronger,
faster, more highly insulated, or which achieve sexual fertility earlier possess a
differential reproductive advantage. Gould declared that

selection simply cannot see genes and pick among them directly. It must use
bodies as an intermediary.[30]

Dawkins would not disagree. He maintained that the selective process

favours some genes rather than others not because of the nature of the genes
themselves, but because of their consequences—their phenotypic effects.[31]

Dawkins has described, in several books,[32] the various ways in which replicators
secure their own survival by increasing the survival-likelihood of the vehicles
that carry them.

On this view of evolution, there is no conflict between "gene selection" and
"individual selection." There is one selection process in which replicator (gene) and
vehicle (individual) play complementary roles.[33] Gould's criticism was off-target.
Whether there is selection at the group level as well remains controversial. Dawkins
was skeptical about the occurrence of natural selection at the group level.[34]

Dawkins promoted the phrase "selfish gene." He maintained that evolution is
a process in which genes strive to increase their chances of survival in the gene

pool. A particular "selfish gene" includes "*all replicas* of a particular bit of DNA."[35] Dawkins noted that what appears to be "altruistic behavior" by an organism instead may be behavior that increases the reproductive chances not of the organism itself, but of copies of its genes in other organisms.

From a biological point of view, "altruism" is measured by the comparative likelihood of reproductive success. There is no question of intentional self-sacrifice. An organism is "biologically altruistic" if its actions increase the reproductive chances of other organisms, but decrease its own reproductive chances.

Among instances of "altruism" noted by biologists are

1. Individual vervet monkeys give calls of alarm to warn their fellow monkeys that predators are near. Alarm calls increase the chances of reproductive success for those warned, but decrease the survival chances of the caller.
2. Young Florida scrub jays help their parents raise additional offspring by providing food and protecting the nest. These "helper birds" sacrifice their own reproductive chances, for a time at least, in order to maximize the reproductive chances of their siblings.
3. Female workers in bee, ant, wasp, and termite societies devote their lives to supplying food for the colony, protecting the nest, and caring for larvae. These female workers do not produce offspring of their own. Instead, they work tirelessly to support the queen of the hive as she produces offspring.

This is puzzling behavior, until one notices the way in which genes are involved in reproduction within the hive. In bee society, for instance, male (drone) bees develop from unfertilized eggs. Each male possesses just one set of genes—those of his mother. Female (worker) bees develop from fertilized eggs. Each female possesses one set of genes from her mother and one set of genes from her father. This "haploidal" reproductive arrangement may be illustrated by tracing the ancestry of a male bee (see Figure 34).

An interesting aspect of this genealogical tree is that the numbers of bees in successive generations conform to the Fibonacci series. The Fibonacci series is a numerical sequence for which each member—after n = 2—is the sum of the previous two numbers of the series, viz., $F_n = [F_{n-1} + F_{n-2}]$, given that $F_1 = 1$ and $F_2 = 1$. The series becomes 1, 1, 2, 3, 5, 8, 13, 21, 34, 55, ... Numerous processes in nature display this pattern, including (1) the axial distribution of leaves as they emerge along the stem of a monocotyledonous plant, (2) the growth of crystals

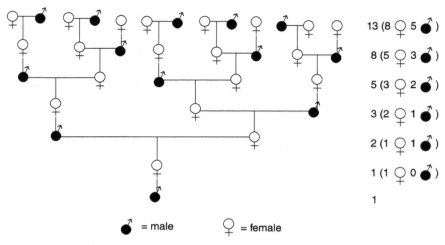

13 (8 ♀ 5 ♂)

8 (5 ♀ 3 ♂)

5 (3 ♀ 2 ♂)

3 (2 ♀ 1 ♂)

2 (1 ♀ 1 ♂)

1 (1 ♀ 0 ♂)

1

♂ = male   ♀ = female

Figure 34 The Genealogical Tree of a Male Bee.

from supersaturated solutions, and (3) the geometry of successive spirals in the growth of a chambered nautilus.

Consider the queen's daughters. They develop from fertilized eggs. Each of these daughters shares 75 percent of her genes with each of her sisters, 100 percent of the genes from her male parent, and 50 percent of the genes from her female parent. If a daughter of the queen were to mate so as to produce female offspring herself, she would share just 50 percent of her genes with each of her daughters.

On the selfish gene theory, the behavior of a worker bee makes sense. She devotes her adult life to increasing the survival chances of her sisters—the offspring that the queen continues to produce. By so doing, the female worker bee maximizes the probability that her genes, 75 percent of which she shares with her sisters, be represented within the hive.

Dawkins noted that, in some cases, the genes of species (1) produce phenotypic changes in species (2). This type of effect often is present in parasite–host relationships. For example, the protozoan parasite *Nosema* manufactures a hormone that prevents the transition of its host beetle from the larval stage to the adult stage. The beetle remains stuck in the larval stage. It continues to grow, however. This is good for the survival and propagation of parasites. It is a disaster for the beetle, since it has lost the ability to reproduce.

A second example is the fluke–snail relationship. Snails that contain the fluke (flatworm) parasite have extra thick shells. This maximizes the snail's survival probability and thereby benefits the fluke. However, it also decreases the likelihood that the snail reproduces, because energy needed to achieve reproductive success has been diverted to shell construction.

A particularly lurid example of parasitic genes causing phenotypic changes in their host is the relationship between the parasite *Sacculina* and the crab upon which it preys. The barnacle-like *Sacculina* sucks nourishment from its host crab. It does so by first destroying the reproductive organs of the crab. The crab responds by diverting energy from activity related to reproduction to a fattening of its own body. This is a process that benefits the parasite. Dawkins declared that

> the changes in the host ... must be seen as extended phenotypic effects of parasitic genes. Genes, then, reach outside their own body to influence phenotypes in other bodies.[36]

Human beings constitute a new stage in the evolutionary process, a stage in which self-conscious awareness has been achieved. The achievement of consciousness has enabled human beings to alter and control their environment in ways unavailable to other species. Dawkins observed that consciousness

> can be thought of as the culmination of an evolutionary trend towards the emancipation of survival machines as executive decision-takers from their ultimate masters, the genes.[37]

Genes remain the "ultimate masters." They determine the way in which the nervous system is constructed. Dawkins maintained that genes are the primary policy makers; brains are

> the executives ... The moment-to-moment decisions about what to do next are taken by the nervous system.[38]

Cultural evolution has been superimposed upon genetic evolution. Dawkins suggested that

> cultural transmission is analogous to genetic transmission in that ... it can give rise to a form of evolution.[39]

The product of biological evolution is a tree of descent. Presumably there is a similar pattern within cultural evolution.

Is there an entity within cultural evolution that plays a role similar to that of genes in biological evolution? Dawkins suggested that there is. He termed this gene-analogue the "meme." The meme, like the gene, is a replicator. It is a "unit of cultural transmission."[40] Just as genes secure their propagation from generation to generation by means of a process of copying, so also memes propagate themselves in the meme pool by leaping from brain to brain via a process which, in the broad sense, can be called imitation.[41]

Candidates for status as memes include scientific ideas, language idioms, songs, oral history traditions, religious ideas and practices, styles of architecture, painting and music, and clothing fashions. Dawkins claimed that

> some memes are more successful in the gene pool than others. This is the analogue of natural selection.[42]

A successful meme has demonstrated adaptation to changing cultural conditions over time. A mark of successful adaptation for a scientific idea is the number of citations recorded in the relevant journals. A mark of successful adaptation for a shoe style is sales statistics from major retailers.

If the evolution analogy is appropriate, there must be vehicles that carry the meme-replicators. We need to fill in the right side of the proportion below:

Gene	:	Animal	::	Meme	:	?
(replicator)		(vehicle)		(replicator)		(vehicle)

What would count as a "cultural vehicle"?

If scientific ideas are memes, then academic departments, professional organizations, and journals are vehicles for their transmission. If religious practices are memes, then congregations of worshippers in churches, synagogues, and mosques serve as vehicles. If musical styles are memes, then performing groups and recordings facilitate their propagation.

There are a number of problems for the gene–meme analogy. The first is the circularity objection that was raised initially against Darwin's theory of organic evolution. Some critics of Darwin's theory complained that it lacked explanatory significance. The theory states that the best-adapted individuals within a population survive. How do we know that selected individuals are well adapted? We establish that they indeed did survive. The critics maintained that there is no independent measure of adaptation over and above actual survival. If the fittest individuals are identified as those individuals that survive, then Darwin's theory tells us that the individuals that survive are the individuals that survive.

Darwin replied that the measure of adaptation is "probability of survival" and not "the fact of survival." He maintained that if (1) an organism possesses a trait that confers an advantage in the performance of some task, and (2) successful completion of this type of task increases the organism's likelihood of survival, then this trait is an adaptive trait. Since empirical evidence can be obtained for (1) and (2), the theory of natural selection does not reduce to a tautology.

A similar objection may be directed at meme theory. There appears to be no way to establish that a meme is fit other than to show that it persists over time. To be fit is to survive. Is there an effective response to this objection? Dawkins suggested that meme-fitness requires copy-fidelity, fecundity, and longevity.[43] Since longevity is persistence over time, it provides no independent measure of fitness. Fecundity is the more promising measure of fitness. Perhaps a proliferation of memes that closely resemble, and yet differ from, an original meme is a measure of the fitness of the original meme. If this is the case, then meme theory is not undermined by the circularity objection.

The second problem is that meme theory has not achieved the conceptual integration required of a high-level scientific theory. Consider the theory of natural selection introduced by Darwin and Wallace. This theory explains the geographical distribution of species, the historical sequence of species revealed in the fossil record, the existence of mimicry, rudimentary and homologous organs, and the adaptation of species to changing ecological conditions.

Meme theory has not achieved a comparable integration of our knowledge of cultural phenomena. It does provide a picture that can be superimposed upon developments in cultural history. For example, we can interpret the career of the idea of heliocentrism as a meme competing for survival by copying itself by "imitation" from the mind of Aristarchus to the minds of Copernicus, Galileo, Kepler, and Newton. Heliocentrism has acquired important relations to other ideas in the course of its evolutionary development. Moreover, the idea has been in competition with other ideas about planetary organization during the course of intellectual history.

We knew all this before there was talk of "meme transmission." It remains to be shown that the notion of memes seeking survival within a meme pool achieves the type of integration achieved by the theories of natural selection, Newtonian mechanics, and the atomic theory.

Other problems for the evolution analogy were pointed out by Dawkins himself. There is an important disanalogy between the interaction of genes and the interaction of memes. In sexual reproduction,

> each gene is competing particularly with its own alleles—rivals for the same chromosomal slot. Memes seem to have nothing equivalent to chromosomes, and nothing equivalent to alleles.[44]

Dawkins downplayed this disanalogy. He suggested that, although a meme does not compete with an allele-meme, it does compete with other memes for attention within the brains of human beings.[45] In addition, there is a competition among memes for exposure in newspapers, on library shelves, and on radio and television.

Dawkins noted that, upon occasion, memes and genes may oppose one another. An example is the meme for celibacy. It has survival value within an interrelated complex of ideas within certain religious groups. Viewed on the level of genes, however, celibacy has no survival value, except in bee, ant, and wasp societies.

Finally, there is a problem about variation in the development of cultural entities. In the case of biological evolution, natural selection operates on a set of antecedently given variants. These variants are not themselves the product of a selection process. They occur randomly and spontaneously through mutation.

The situation is quite different in the development of cultural entities. The variations on which the process of "cultural selection" operates are not generated randomly. The competition involved in the selection process typically is between lineages of ideas, styles, or practices. Each idea or practice in a lineage is a modification of its predecessor.

This competition is markedly different from natural selection operating on a collection of giraffes of varying neck length under conditions of scarcity of low-lying vegetation, or a collection of finches of varying beak shapes in an environment in which nuts are the most available food supply.

Given this collection of problems, it would seem futile to seek to develop an analogy between biological evolution and cultural evolution. Nevertheless, a number of scholars sought to interpret aspects of cultural development to be products of a process of natural selection.

In *The Extended Phenotype* (1992) Dawkins acknowledged that he had misidentified memes in *The Selfish Gene* (1976). His new position in *The Extended Phenotype* is that a meme-replicator is

> a unit of information residing in a brain ... It has definite structure, realized in whatever physical medium the brain uses for storing information.[46]

The ideas, styles, and practices Dawkins had taken to be memes in *The Selfish Gene* he now recognized to be "meme-products" or "phenotypic effects."[47] Architectural styles, fashions, catchy tunes, and the competing ideas of heliocentrism and geocentrism are meme-products and not memes themselves. Dawkins' revised analogy between biological evolution and cultural evolution is based on the following table:

Replicator	Replicator-Product	Vehicle
Gene	Phenotype e.g., long neck, thick beak	Organism e.g., giraffe, finch
Meme	Meme-product e.g., ideas, styles, practices	Meme-vehicle e.g., individual human beings, groups of human beings

A gene replicates by means of a copying process. A great deal is known about the details of this process. On the meme–gene analogy, there should be a similar copying process for the transmission of memes—*qua* "unit of information residing in a brain"—from vehicle to vehicle.

Meme transmission presumably takes place when brain-states are modified in certain ways. Detailed knowledge of such brain-state changes is not available. Nevertheless, it is clear that certain features of meme transmission differ from those of gene transmission. Dawkins conceded that there are important differences between memes—on this revised interpretation—and genes:

1. "Memes are not strung out along linear chromosomes, and it is not clear that they occupy and compete for discrete 'loci,' or that they have identifiable 'alleles'";[48]
2. "The copying process is probably much less precise than in the case of genes";[49]
3. "Memes may partially blend with each other in a way that genes do not";[50] and
4. "New 'mutations' [of memes] may be 'directed' rather than random."[51]

Dawkins drew a somber conclusion from these disanalogies. He declared that

> these differences may prove sufficient to render the analogy with genetic natural selection worthless or even positively misleading.[52]

The program to show that cultural change is analogous to biological evolution has not achieved success. Cultural change is not the product of natural selection acting upon a population of antecedently given cultural variants.

Dawkins began *The Extended Phenotype* with a disclaimer about its content. The book does not put forward a scientific theory for which confirming or disconfirming evidence can be accumulated. Instead, it develops a theoretical framework for the interpretation of the progression of life forms during the history of the Earth. The theoretical framework is based on the idea that the replicator should be thought of as having *extended* phenotypic effects, consisting of all its effects on the world at large, not just its effects on the individual body in which it happens to be sitting.[53]

Dawkins maintained that one role of the gene–meme analogy is to reinforce and extend this interpretation.[54] He declared that the main value of the gene–meme analogy

> may lie not so much in helping us to understand human culture as in sharpening our perception of genetic natural selection.[55]

Genes and chromosomal bits are the primary replicators. The meme presumably also is a replicator. Meme-products clearly influence vehicles other than the vehicles responsible for their production.

It might seem that a successful meme theory would provide support for the extended-phenotype view that Dawkins sought to promote. Unfortunately, as noted above, Dawkins himself admitted that the gene–meme analogy may be "worthless or even positively misleading." The fact that meme-products, like phenotypes, are "extended" in their influence does not establish that cultural change and biological evolution are analogous processes.

One senses that it is with regret that Dawkins concluded that cultural evolution is not an analog of biological evolution. Joseph Fracchia and Richard Lewontin, by contrast, appeared to celebrate the failure of the evolution analogy.[56]

Fracchia and Lewontin insisted that an effective analogy based on Darwinian evolutionary theory must develop cultural equivalents for (1) random variation, (2) inheritance, and (3) natural selection that produces differential reproduction rates.

Memes are popular candidates for objects subject to random variation within cultural evolution. However, Dawkins was correct to admit that memes are not randomly generated. He suggested that their mutations may be "directed."[57] Biological evolution is based on uncoupled processes of variation and selection. Without a cultural analog of random mutation, the first requirement for an effective meme–gene analogy is not fulfilled. Fracchia and Lewontin posed a question that meme-theorists have failed to answer:

> For a variational theory, it must be possible [to] count up the number of times each variant is represented. What is the equivalent for memes of the number of gene copies in a population?[58]

Inheritance is the second requirement. Meme-theorists have no difficulty providing examples of meme-product (e.g., ideas, practices) transmission. For meme-product transmission to count as an analogy of biological inheritance, however, a meme—*qua* "unit of information residing in a brain"—must be subject to a copying process that preserves it from a generation (1) vehicle to a generation (2) vehicle. This requisite copying process has not been established.

Moreover, as Fracchia and Lewontin suggested, it may be more appropriate to speak of cultural forms as "acquired" rather than "transmitted" or "inherited." They declared that

> acculturation occurs through a process of constant immersion of each person in a sea of cultural phenomena, smells, tastes, postures, the appearance of buildings, the rise and fall of spoken utterances.[59]

This "sea of cultural phenomena" is mediated through family, teachers, social and institutional groups, radio, television, and books. Fracchia and Lewontin concluded that the complexity of the modes of acquisition of cultural forms makes unlikely the success of an analogy to biological inheritance.

Natural selection is the third requirement for an effective evolutionary analogy. Some meme-products survive over time; others do not. One may identify cultural forms whose histories resemble descent with modification, but the existence of such patterns does not pinpoint a mechanism responsible for them. The cultural theorist may superimpose a concept of "survival of the fittest" upon the development of cultural forms, but, as Fracchia and Lewontin conclude,

> it is unclear what useful work is done by substituting the metaphor of evolution for history.[60]

L. J. Cohen on the Inappropriateness of the Evolution Analogy

Unfortunately for the evolution-analogy program, there are important disanalogies between organic evolution and the growth of science. L. J. Cohen pointed out two such disanalogies.

In the first place, the process by which variants are produced within a breeding population takes place independently of the process by which the "better-adapted" individuals succeed in the struggle to survive and reproduce. Mutation is a spontaneous, random process. Natural selection operates on a pool of antecedently given variants. The result is a differential reproductive advantage for those individuals best adapted to overcome challenges posed by the environment.

Cohen emphasized that the *generation* of variants and the *selection* of variants are distinct, "uncoupled" processes. He contrasted these processes in the following colorful passage:

> The gamete has no clairvoyant capacity to mutate preferentially in directions preadapted to the novel ecological demands which the resulting adult organisms are going to encounter at some later time.[61]

In the second place, biological species are not analogues of scientific disciplines. Nor are biological species analogues of "scientific research programmes" that become implemented in sequences of theories (Lakatos). A biological species is a population of similar individuals, each of which is a representative of that species. The same is not the case for scientific research programs. Scientific research programs include concepts, invariant and/or statistical relations among concepts, theories about underlying mechanisms, procedural rules, and evaluative standards. These diverse ingredients are interrelated in complex ways.

The identity-through-change of a biological species is markedly dissimilar to the identity-through-change of a scientific research program. A biological species retains its identity provided that a set of individuals with similar characteristics at time t_2 resembles in relevant respects another set of individuals with similar characteristics at time t_1. But the identity-through-change of a scientific research program is not of this type. In order to solve conceptual problems within a research program we need a set of interrelated concepts, not a population of concepts with similar characteristics. Hence changes within a research program involve a restructuring of an "evolving" concept's relations to other concepts and not just a replacement of concepts similar to C_1 by concepts similar to C_2.

Consider, for example, the "struggle for existence" of two concepts in eighteenth-century chemistry—"phlogiston" and "oxygen." There were competing theories about the burning of metals. Joseph Priestley (1733–1804) interpreted the burning of metals as a release of phlogiston, e.g.,

Zn → calx of zinc + phlogiston ↑ (calx of zinc + phlogiston).

Antoine Lavoisier (1743–1794) interpreted this process as a combination of zinc with oxygen. In modern notation

$2 \, Zn + O_2 = 2 \, ZnO$ (calx of zinc)

Lavoisier's oxygen theory prevailed. One may view the transition from phlogiston theory to oxygen theory as an episode in which the better-adapted concept was selected. However, there is no species instantiated by phlogiston.

Moreover, scientific concepts such as "electron," "heat," "force," and "gene" fail to display the identity-over-time possessed by biological species such as horses and elephants. The evolutionary analogy does not apply to scientific concepts.

The adequacy of an evolutionary-analogy theory of science depends on the importance of the above-mentioned disanalogies. Cohen insisted that the independence of variation-generation and selection is an *essential* feature of the theory of natural selection. He concluded that the analogy to the growth of science fails. Toulmin and Hull, by contrast, conceded that this disanalogy exists, but maintained that the evolutionary analogy nevertheless provides a useful theory of science.

Donald Campbell on Blind Variation and Selective Retention

Donald Campbell sought to reinstate the evolutionary analogy by shifting attention from "random mutations" to "blind trials." He acknowledged that scientific beliefs, unlike biological variants, are not produced randomly. The scientist has in mind a problem to be solved and a history of prior attempts to find a solution. However, if scientific progress results from the selective retention of blind trials, then the core of the evolutionary analogy may be retained. Every random trial is a blind trial, but a trial may be blind without being random.

In Campbell's usage, a "blind trial" is a trial that satisfies three conditions: (1) the trial is independent of environmental conditions; (2) a successful trial is no more likely to occur at one point in a series of trials than at any other point in the series, and; (3) no trial in a sequence of trials is put forward as a "correction" of a prior trial. Campbell claimed that

> a blind-variation-and-selective-retention process is fundamental to all inductive achievements, to all genuine increases in knowledge, to all increases in fit of system to environment.[62]

Campbell thus maintained that the goal-directed decisions of scientists to entertain specific hypotheses are "blind" forays into the unknown, and that those hypotheses that prove "non-adaptive" under testing are eliminated.

The "selective-retention-of-blind-variants" view, like Thagard's computational philosophy of science, is a program for further research. In order for Campbell's program to succeed as a robust descriptive philosophy of science, two conditions must be fulfilled. The first condition is that the evolutionary analogy—amended to require selective retention of "blind" variants—must fit important episodes from the history of science. The second condition is that the "fit" has explanatory force.

Kepler's work on the orbit of Mars is a promising candidate for the blind-trials theory of variant-concept creation. Kepler superimposed a variety of ovoid orbits on the data before settling on an elliptical orbit with the sun at one focus. Are the three conditions for blind trials fulfilled in Kepler's work?

Condition (1) requires that a blind trial be independent of the environment conditions of the occasion of their occurrence. It is unclear how to establish this independence. Kepler worked with a database of successive positions of Mars along the zodiac. This database was an essential part of the problem-situation that Kepler sought to solve. The selection of a particular ovoid orbit clearly was not independent of this database. If the trial was "blind" then the database was not part of the environmental conditions present on the occasion of its occurrence. The relevant environmental conditions remain to be specified.

Condition (2) requires that a specific correct trial be no more likely to occur at any one point in a series of trials than at another point. The superposition of an ellipse is the "correct" trial in this series. I believe that most historians of science would hold that Kepler was more likely to select an ellipse at the end of the series of ovoid-orbit trials than at the beginning. If so, then Kepler's ellipse trial was not a blind trial.

Condition (3) states that a variation subsequent to an incorrect trial is not a correction of that trial. But surely the ellipse variation is a correction of the last ovoid-orbit variation in the sequence. Scientists at the time judged the ellipse variation to have the status of a law, a status not accorded to any of the preceding ovoid variations.

If the blind-trials approach fails to reproduce this *prima facie* favorable instance of variant-concept creation, its prospect as a rational reconstruction of scientific progress is not promising. Moreover, the creation of such concepts as "entropy," "neutron," and "superego" present more formidable obstacles to the blind-trials theory of concept creation.

Moreover, Campbell's modified evolutionary analogy is not promising as an account of theories about underlying mechanisms. It is implausible that Descartes' vortex theory, the kinetic theory of gases, molecular orbital theory, and plate tectonics theory are results of the selective retention of blind trials.

Campbell's "blind-variation-and-selective-retention" theory of scientific change would appear to be subject to two principal challenges. In the first place, the generation of variant hypotheses is neither random nor blind. Rather, hypotheses are put forward in response to recognized inadequacies within the scientific environment. In the second place, the sorting process that results in a decision to reject a high-level hypothesis involves judgments about background knowledge, auxiliary assumptions, experimental procedures, and sometimes even metaphysical principles. Ron Amundson declared that

> the process of rejecting a "falsified" hypothesis is often more like the literary critic's negative assessment of a poem than like the cold wind's freezing of the baldest polar bear.[63]

Campbell sought to blunt this type of criticism by acknowledging the existence of "shortcuts" for the blind-variations-and-selection process. The formulation of hypotheses designed to address perceived deficiencies within a scientific domain may serve as a shortcut for the method of blind trials. Campbell insisted that such shortcuts themselves are an inductive achievement. He declared that we have acquired a

> wisdom about the environment achieved originally by blind variation and selective retention.[64]

Purposefully generated (nonblind) hypotheses may be part of a "phylogenetic lineage" the earlier members of which were generated by the blind-variation-and-selective-retention process.

Campbell's introduction of "shortcuts" raises anew the question about the explanatory force of his modified evolutionary analogy. Does it have explanatory force to argue that scientists entertain goal-directed hypotheses today because at sometime in the past our ancestors engaged in blind trials (upon which selection operated)?

Popper on Conjectures, Refutations, and the Evolutionary Analogy

Karl Popper endorsed Campbell's version of the evolutionary analogy. He held that the conjectures of scientists are analogous to variations and that refutations are analogous to the selective retention of variants. On this view, the "phylogeny" of science is the lineage of theories that survives the rigors of severe testing.

Popper's opposition to inductivism is well known. He repeatedly insisted that there can be no successful algorithm for theory-formation. Popper likened the position of the theorist to the

> situation of a blind man who searches in a dark room for a black hat which is— perhaps—not there.[65]

The theorist, like the blind man, proceeds by trial-and-error, coming to learn where the hat is not, without ever reaching a certainty immune from rejection from the force of further experience.

Popper is correct to emphasize the role of creative imagination in the formulation of scientific hypotheses. The problem-situation does not dictate a solution to the theorist. However, neither are hypotheses formulated independently of the problem-situation. Popper's "black hat" image is quite misleading. Scientific conjectures are "blind" only in the sense that the outcome of subsequent testing is unknown. They are not "blind" in Campbell's sense of being "independent of the environmental conditions of the occasion of their occurrence."[66]

There is a further difficulty in Popper's particular use of an evolutionary analogy. Popper insisted that scientists ought formulate bold, content-increasing conjectures that run a high risk of falsification. But the Darwinian picture of descent with modification is a gradual accretion of small adaptations. Popper claimed explanatory value for an "evolutionary analogy" that includes pious references to "Darwinian theory." But he also introduced "Lamarckian" emphases and "saltation" effects that are inconsistent with that theory.

Of course, there have been disputes over the specific content of Darwinian theory. Nevertheless, Michael Ruse is correct to conclude that

> Popper has been no more successful than others in making traditional evolutionary epistemology plausible. The growth of science is not genuinely Darwinian.[67]

The Evolutionary-Origins View

The evolutionary-origins view, like the evolutionary-analogy view, may be defended as a purely descriptive theory about science. The evolutionary-analogy view is that there is competition leading to differential reproductive success within both organic evolution and science. The evolutionary-origins view is that scientific inquiry is directed by the application of epigenetic rules that have become encoded in *homo sapiens* in the course of evolutionary adaptation. We have certain capacities and dispositions because it was advantageous for our ancestors to have them.

Michael Ruse on Epigenetic Rules

Michael Ruse called attention to several epigenetic rules that appear to inform human evolution: (1) the partitioning of the (continuous) spectrum into discrete colors that takes place in diverse human cultures, presumably because it conferred adaptive advantage in the struggle for existence; (2) the "deep structure" of language uncovered by Chomsky and others; and (3) the prohibition of incest.

Ruse suggested that there exist additional epigenetic rules that govern the creation of science: (1) formulate theories that are internally consistent; (2) utilize the principles of logic and mathematics in the formulation and evaluation of theories; (3) develop theories that are "consilient" (Whewell); and (4) seek "severe tests" of theories (Popper).[1]

The first rule is beyond criticism. Internal consistency is a necessary condition of cognitively significant discourse in general, and science in particular.

The second methodological rule likewise might seem to be beyond criticism, but its status is more controversial. Emphasis on mathematics can be overdone.

Pythagoreanism is an extreme position on the role of mathematics in science. The committed Pythagorean believes that there are mathematical harmonies in the universe and that the role of the scientist is to uncover these harmonies. To show that particular types of phenomena are mathematically related is to understand why the phenomena are what they are.

Johannes Kepler was committed to the Pythagorean orientation. He noted with pride that there is a mathematical relation between the densities of important substances found on earth and the mean distances of the known planets from the sun. These densities are inversely proportional to the square root of the planetary distances (see Figure 35).

The "densities" calculated from planetary distances are in close agreement with the measured densities of the corresponding terrestrial substances. The appropriate reaction to Kepler's relation is "so what?" The search for mathematical relation of this type does not contribute to the progress of science.

Planet	Substance	Density = $1 / \sqrt{\text{distance}}$ (Earth = 1000)
Saturn	Diamond	324
Jupiter	Lodestone (iron oxide)	438
Mars	Iron	810
Earth	Silver	1000
Venus	Lead	1175
Mercury	Quicksilver (Mercury)	1605

Figure 35 Kepler's Density–Distance Relation.

Source: Johannes Kepler, *Epitome of Copernican Astronomy*, trans. C. G. Wallis, in *Ptolemy, Copernicus, Kepler: Great Books of the Western World*, vol. 16 (Chicago: Encyclopedia Britannica, 1952), 882.

In other instances, however, Kepler's Pythagorean commitment paid dividends. He was convinced that there must be a mathematical relation between the distances of the planets from the sun and the times required for a revolution around the sun. He believed that God created the solar system on a mathematical model. Kepler tried various mathematical relations and finally found a relation that fit observations. The third law of planetary motions states that $(T_1/T_2)^2 = (d_1/d_2)^3$, where T_x is the period of planet x and d_x is its mean distance from the sun.

Since the third law holds for the earth, considered to be a planet, this relation provided important support for the sun-centered theory of the solar system. Is this relation an application of mathematics to express a theory? Does it count as a scientific theory that one believes that God is a mathematician? Clearly not. However, Kepler's period–distance relation subsequently was incorporated into Newton's theory of the solar system. The period–distance relation is a deductive consequence of Newton's three axioms of motion and the law of universal gravitational attraction on the assumption that the masses of the planets are negligible compared with the mass of the sun. Historians of science are unanimous that Kepler made significant contributions to scientific progress by seeking mathematical relations everywhere, despite the irrelevance of some of those relations.

The same cannot be said for Josiah Cooke's search for mathematical relations among the atomic weights of the chemical elements. Cooke (1827–1894) was a prominent chemist who taught for decades at Harvard University. Cooke examined groups of chemical elements the members of which display similar chemical properties. One such group is the halogens. Each member of the halogen group—F, Cl, Br, I—forms a compound with sodium of the form NaX. Cooke believed that there must be some mathematical harmony that accounts for this. He observed that the atomic weights of the halogens satisfy the relation $W = 8 + 9\,n$, where W is the atomic weight and n is an integer. He exhibited this relation as shown in Figure 36.

The observed atomic weights, determined at the time, were in close agreement with the values calculated from selected values of n. But why take values of n to be 1, 3, 8, and 13? This seems arbitrary. Why not $n = 4$ or $n = 11$? Are we to search for additional halogen elements whose weights correspond to these values of n?

Cooke's preoccupation with algebraic relations even led him to group together elements with disparate chemical properties. Working with erroneous data that attributed atomic weight 6 to carbon and 21 to silicon, Cooke noted that the set of elements carbon, boron, and silicon have atomic weights that conform to the relation $W = 6 + 5\,n$.

element	F	C1	Br	In
n	1	3	8	13
atomic weight (calculated)	17	35	80	125

Figure 36 Cooke on the Halogens.

Source: Josiah P. Cooke, "The Numerical Relation between the Atomic Weights, with Some Thoughts on the Classification of the Chemical Elements," *American Journal of Science 17* (1854), 405.

element	C	B	Si
n	0	1	3
atomic weight (calculated)	6	11	21

Figure 37 Cooke on the Carbon-Boron-Silicon Series.

Source: Cooke, "Numerical Relation," 406.

Carbon and silicon have similar properties, but the chemical properties of boron are quite different. In fairness to Cooke, it should be noted that there was little reliable knowledge about boron's compounds available at the time. Nevertheless, his applications of mathematical formulas to chemical phenomena did not contribute to the progress of science.

Benjamin Brodie (1817–1880) promoted an even more extensive program to explain chemical reactions by reference to algebraic relations. Brodie was professor of chemistry at Oxford University. He was active in the Chemical Society of London, serving as both its secretary and its president.

Brodie initiated a "chemical algebra" by taking the symbol α to represent the operation by which a unit of space is converted into a unit of hydrogen. He then took the symbol χ to represent the conversion of a unit of hydrogenated space into a unit of hydrogen chloride.[2] He then represented hydrogen chloride as the product of successive operations α and χ. The decomposition of hydrogen chloride then becomes

$$2\,\alpha\,\chi = \alpha + \alpha\,\chi^2$$

hydrogen chloride = hydrogen + chlorine

He obtained similar results for bromine ($\alpha\,\beta^2$), viz.

$$2\,\alpha\,\beta \qquad = \qquad \alpha \qquad + \qquad \alpha\,\beta^2$$

| hydrogen bromide | | hydrogen | | bromine |

and for nitrogen

$$2\,\alpha^2\,\upsilon \qquad = \qquad 3\,\alpha \qquad + \qquad \alpha\,\upsilon^2$$

| ammonia | | hydrogen | | nitrogen |

where β is the operation of converting a unit of hydrogenated space into a unit of hydrogen bromide, and υ is the operation of converting a unit of hydrogenated space into a unit of ammonia.

These results suggested to Brodie that chlorine ($\alpha\,\chi^2$), bromine ($\alpha\,\beta^2$) and nitrogen ($\alpha\,\upsilon^2$) are compounds of hydrogen and yet-to-be-discovered elements whose operations are χ, β, and υ. He recommended that chemists look for and identify these elements.

Some scientists accepted the challenge. In 1870, Brodie's former research associate Norman Lockyear discovered a hitherto unrecognized element in the spectrum of the sun. This new element was named "helium." At that time, helium had not been found on the earth.

This discovery gave temporary encouragement to scientists who sought to implement Brodie's program. However, further results were not forthcoming. By the end of the nineteenth century Brodie's chemical algebra was of interest only to historians of chemistry. As a research program it no longer had followers.

Brodie's chemical algebra was not a completely non-adaptive application of mathematics in support of a theory. It did provide a rationale for the search for new elements in the spectrum of the sun. On balance, however, Brodie's chemical algebra proved to be a dead end. It failed to yield important new discoveries, and it diverted research away from more fertile research programs.

The most effective criticism of the role of mathematics in science is that applications of mathematics are too easy to make. Given a set of data points relating types of phenomena A and B, a competent mathematician can create a power expansion of the form $A = k + aB + bB^2 + cB^3 + \ldots$ that fits the data. The closeness of the fit always can be increased by increasing the number of terms in the power expansion.

In general, power expansions, and the engineer's "rules of thumb" do not contribute to scientific progress. The use of mathematics to express a theory is another matter. The contrast between these two uses of mathematics may be illustrated by reference to relations of classical thermodynamics.

The virial equation—$P V + k T + aT/V + bT/V^2 + cT/V^3 + \ldots$—provides superior ability to reproduce observed pressure-volume-temperature data. The constants $a, b, c, \ldots$ are determined empirically for the particular gas under study. The fit with observed data can be increased by increasing the number of (xT/V^n) terms employed.

The ideal gas law—$P V = n R T$—provides a less accurate fit to P-V-T data. This is particularly the case at high pressures. However, the ideal gas law contributes to our understanding of thermodynamic processes. It can be derived from an idealized version of kinetic molecular theory in which molecules are treated as point-particles. The kinetic theory provides an explanation for the expansion of a balloon with increasing temperature. The higher the temperature, the more frequent and more energetic are the collisions of gas molecules with the inner surface of the balloon. The virial equation provides no such explanation.

Overall, applications of mathematics have made significant contributions to scientific progress. Ruse's second methodological rule, properly qualified to exclude uses of mathematics that merely "save appearances," is a suitable candidate for status as an epigenetic rule.

The third epigenetic rule was championed by William Whewell (1794–1866). According to Whewell, a theory achieves consilience provided that

1. it is logically consistent;
2. it is more inclusive than its predecessor; the range of facts it accounts for is greater than the range of facts accounted for by its predecessor; and
3. this increase in explanatory power is accompanied by a gain in simplicity.

Whewell emphasized the vast consilience achieved by Isaac Newton by means of the three axioms of motion and the law of gravitational attraction. Newtonian mechanics incorporates Kepler's laws of planetary motion, Huygens' theory of momentum transfer in collisions, the motions of pendulums and the tides, and simple harmonic motion. In addition, Newton's theory provides explanations for

1. irregularities in the motion of the moon;
2. perturbations in the orbits of Jupiter and Saturn caused by their mutual gravitational attraction;
3. the precession of the equinoxes; and
4. the decrease in weight of bodies moved from the earth's pole to the earth's equator.

It does so in a way that satisfies the requirement of a gain in simplicity. Newton's mechanics subsumes empirical relations under the ideas of inertial motion, universal gravitational attraction, and the direct proportionality of force and acceleration.

If one were inclined to quibble, one could highlight an occasional example of scientific progress in which the methodological rule to achieve consilience is disobeyed. For instance, Galileo contributed to scientific progress by *separating* kinematics from the general study of the motions of bodies. In the analysis of freely falling bodies, he put aside questions about the causes, or forces, responsible for motion and focused exclusively on the variables position, time, velocity, and acceleration. This anti-consilient approach proved fruitful. Galileo demonstrated that a body in free-fall travels distances 1, 3, 5, 7, 9, ... in successive equal time intervals. This "law of odd numbers," an equivalent formulation of which is that the acceleration of a falling body is constant, was established without consideration of the cause of the motion. Galileo evidently was not genetically hard-wired always to seek consilience.

Progress also is sometimes achieved by disregarding the fourth methodological rule, which requires acceptance of the results of a "severe test." As noted above (p. 167), Einstein disregarded the apparently decisive disconfirmation of the theory of special relativity by an experiment of Walter Kaufmann in 1906. Einstein took the position that the theory of special relativity must be true and that Kaufmann's *prima facie* experimental refutation of the theory must be incorrect. In retrospect it is clear that Einstein was right to disregard Kaufmann's test result and to continue to develop the implications of the theory of special relativity. Gerald Holton emphasized that Einstein could not have known that Kaufmann's experimental procedure was inadequate. The inadequacy was established only ten years later.[3]

Einstein evidently was not genetically hard-wired always to accept refutation by a severe test. Of course, it may be the case that the majority of successful scientists are genetically hard-wired to pursue consilience and the severe tests of theories, despite the fact that some scientists are not wholly committed to these methodological directives.

The methodological principles Ruse held to be hard-wired epigenetic rules are quite general. Despite the exceptions noted above, an extensive list of episodes can be compiled for which implementation of these rules has contributed to scientific progress.

More specific methodological principles have not fared well in the history of science. Some formerly accepted methodological rules that subsequently have been abandoned are:

1. A complete scientific explanation of a process must include reference to the *telos* of the process. It must include a statement of the form "*x* occurred in order that *y* be realized" (Aristotle).
2. The cause of every motion is pressure or impact on the moving body (Descartes).
3. Life processes must be explained in terms of the operation of "vital forces" (Whewell).
4. Seek to establish conservation principles. Conservation of mass and conservation of parity have been abandoned. Conservation of mass-energy survives.
5. That which is localizable cannot be said to have wave-like properties, and that which cannot be localized cannot be said to have particle-like properties. These principles have been abandoned in favor of wave–particle dualism.
6. A complete explanation of a physical process must specify a causal relationship that is applicable to the continuous spatio-temporal unfolding of the process. This principle has been abandoned and replaced by a principle of complementarity, at least within the Copenhagen interpretation of quantum mechanics.

Critics of the evolutionary-origins view have pointed out that there is evidence that human beings often make decisions that are inconsistent with these supposedly "genetically hard-wired" rules. Human subjects affirm the consequent with impunity, succumb to the "gambler's fallacy," and erroneously conclude that the probability of (A & B) is higher than the probability of *A* alone. Ruse acknowledged that this is evidence against the evolutionary-origins view, but insisted that it is

> better surely to suppose that much of the time we do not think particularly carefully or logically simply because it is not really necessary to do so, but when pressed we can do so and for very good reasons, namely, that those who could not tended not to survive and reproduce.[4]

This is unconvincing. If certain dispositions are acquired in the evolutionary process because of their adaptive value, then these dispositions ought to be uniformly actualized. Ruse is forced to subdivide human actions into those that conform to the epigenetic rules (performed by scientists) and those that do not conform to those rules (performed by nonscientists in cases where it is "not really necessary" to conform). Ruse does not argue that those who fail to apply the epigenetic rules are likely to succumb to evolutionary pressures. Instead he introduces the *ad hoc* hypothesis that nonconformity occurs in cases in which conformity is not necessary. It is debatable whether such a move is consistent with the empirical method required by the position of evolutionary naturalism.

Normative-prescriptive content may be added to the evolutionary-origins view by endorsing the move from

1. the application of methodological rule *R* and evaluative standard *S* were adaptive responses to former ecological pressures *E*, to
2. *R* and *S ought be* applied by scientists today.

Ruse usually resisted the temptation to make this move. However, he sometimes teased the reader with suggestive adaptive scenarios. For example, he contrasted the responses of two hominids to evidence of the presence of tigers. Hominid (1) takes the existence of feathers, blood, paw marks in the mud, and growls from the bushes to establish a consilience of inductions, and flees. Hominid (2) views the same evidence but fails to see the importance of consilience. Ruse then asked "which one of these was more likely to be your ancestor?"[5] Ruse has led the reader to the point where it is natural to conclude not only that the disposition to apply a standard of consilience has adaptive value, but also that this standard ought be applied within science today.

However, Ruse did not issue explicit normative-prescriptive claims on behalf of epigenetic rules. Instead, he compared epigenetic rules to David Hume's "dispositions." Hume had observed that we organize our lives by reading "necessary connections" into nature. We act on the expectation that correlations experienced in the past will continue to hold in the future. Ruse accepted Hume's account of the dispositions involved in our commerce with the world and appended to the Humean account a theory about the origin of these dispositions.

Hume denied that a rational justification can be provided for our expectations of regularity. Past uniformity does not entail the future continuation of that uniformity. Ruse accepted this Humean claim as well. He suggested that the only "justification" for implementing epigenetic rules is that these rules did arise during the course of human evolution. Ruse acknowledged that to provide a theory about the origin of a rule is to fall short of providing a justification for continuing to implement the rule.

There is a further difficulty for normative-prescriptive versions of evolutionary naturalism. In biological evolution, "fitness" is a balance between a successful adaptation to present environmental conditions and the retention of the capacity to respond creatively to future changes in those conditions. In a particular case, successful adaptation may be achieved at the expense of a loss of adaptability. That this has occurred becomes evident only with the passage of time.

Given a specific evaluative situation, the evolutionary naturalist stipulates that a particular decision is correct only if it promotes fitness in the long run. But how can one know at the time a decision is made that it will do so? One may appeal to the fact that similar decisions in the past have proved to have survival value. However, it is always possible that the ecologically unique present situation requires a different decision.

The most adequate appraisals are those rendered long after the fact. Survival is the best indicator of fitness. It is survival that establishes a continuing retention of adaptive capacity in the face of changing conditions. The major conceptual innovations of Newton and Einstein pass this test. These innovations participate in lineages to which subsequent scientists have contributed. Judgments about contemporary conceptual changes are much less secure.

This is not an objection to evolutionary naturalism as a descriptive philosophy of science. Scientific theories are subject to modification and replacement, and so too are the judgments issued by the philosopher of science. What initially appeared to be a "fit" response may turn out subsequently not to be such. It is no more realistic to expect certainty in philosophy of science than it is to expect certainty in science.

The Strong Programme

Evolutionary-origins view is a causal analysis of scientific evaluative practice. The causes in question are remote, however. The "strong programme" of David Bloor, Barry Barnes, Stevin Shapin, and others is an attempt to uncover the proximate causes of scientists' evaluative decisions. The strong programme unfolds in two stages: (1) description of scientific evaluative practice in its socio-economic context, and (2) analysis of the causes of individual evaluative decisions.

David Bloor listed four presuppositions of the program:

1. the aim of the program is to discover the causes of scientists' beliefs and decisions;
2. causal analyses ought be given for both successful and unsuccessful developments in science;
3. the same types of cause are to be invoked to account for both successful and unsuccessful developments; and

4. the interpretations formulated within the program are themselves subject to causal analysis.[1]

The strong programme was developed in opposition to that approach to the history of science that consigns psychological and sociological investigations to "nonrational" aberrations. Lakatos, for instance, had subdivided the history of science into "internal" developments that conform to appropriate standards of rationality and "external" developments not subject to rational reconstruction. Supporters of the strong programme insisted that the same type of causal analysis be given for successful and unsuccessful scientific investigations. The focus of their causal analyses was the underlying social context within which science is created.

This social context is given structure by the goals and interests that inform research. According to Barnes, goals and interests

> operate as contributing causes of the actions or series of actions which constitute the research . . . it is not knowledge which is treated as caused, but the *change in knowledge* brought about by an action. Goals and interests help to explain the change as the consequence of a goal-oriented or interested action.[2]

Supporters of the strong programme admitted that the goals and interests shared by groups of scientists are not sufficient conditions of individual evaluative decisions. However, they maintained that no causal analysis is complete that ignores these "contributory causes." Bloor cited Shapin and Schaffer's *Leviathan and the Air-Pump*[3] and Rudwick's *The Great Devonian Controversy*[4] as examples of case studies that show the importance of social pressures on the development of science. Bloor, together with Barry Barnes and John Henry, contributed an analysis of R. A. Millikan's data selection in his debate with Felix Ehrenhaft about the alleged minimal unit of electronic charge.[5]

Millikan divided his observations of the motions of oil droplets into those to be accepted for purposes of calculation and those to be discarded. Many of the rejected observations, if used to calculate a value of electron charge, would have yielded values much lower than the value derived from the accepted observations. Inclusion of the rejected observations would have weakened Millikan's thesis—criticized by Ehrenhaft—that each electron has the same indivisible charge. Bloor, Barnes, and Henry concluded that

> The repertoire of interpretive processes used by Millikan depended on a local cultural tradition.[6]

This tradition warranted inferential risk-taking on behalf of a commitment to atomism.

The adequacy of the strong programme as a robust descriptive philosophy of science depends on what is claimed for the program. Michael Friedman observed that the strong programme superimposes a philosophical agenda upon its socio-causal analyses of scientific evaluative practice.[7] The strong programme has two principal goals: (1) the creation of an empirical scientific discipline that uncovers the causes of the evaluative decisions of scientists, and (2) the repudiation of the ideal of inviolable universal standards of truth, rationality, and progress.

Commitment to these two goals generates tensions within the program. Supporters of the strong programme have maintained that there are no standards of truth and rationality apart from collective practices within local, particular contexts. Bloor, for instance, speaks of knowledge as that which is "collectively endorsed."[8] But since reflexivity also is a tenet of the strong programme, epistemological relativism applies as well to interpretations developed within the program. The Barnes-Bloor-Henry interpretation of Millikan's experimental research on electronic charge, for example, is "true" ("accurate," "cogent" …) if, and only if, it is endorsed by interpreters of science. There is a problem, of course, about the membership of the appraisal-community. Is it to be restricted to those who support the strong programme, or should it include those interpreters of science who are skeptical about a "social construction" of knowledge?

If one considers not its philosophical agenda, but the goal of a socio-causal interpretation of science, then the strong programme has achieved several successes. Recent context-sensitive studies of the achievements of Galileo, Boyle, Murchison, and high-energy particle physicists have contributed to our historical understanding.[9] These studies support the strong programme position that interpretation of the evaluative decisions of scientists must take into account the social context within which the decisions are made.

There remains a difficulty about the identification of social causes. To claim that something is a "contributing cause" of an effect is to claim that the effect would not have been what it was in its absence. In historical interpretation, there is no possibility of a replay in the absence of the putative cause. The most that can be done for a claim of a "contributory social cause" is to show that in other historical instances in which the social factor is absent, no effect similar to the one under consideration took place. For instance, one might undertake to show that Newton's preoccupation with alchemy is correlated with social pressures that did not affect his rivals. Or perhaps one might undertake to trace Galileo's program to extend "divine" uniform circular motions to the terrestrial realm to social pressures which affected him but not his rivals. In such cases, the "effects" to be explained are commitments to "Holtonian themata" rather than individual evaluative decisions. It seems clear that many *individual* evaluative decisions are caused by beliefs about reasons. For example, Newton rejected Descartes' vortex theory because he believed that the vortex theory was inconsistent with Kepler's laws. And Davy rejected Lavoisier's oxygen theory of acids because he believed that muriatic acid (hydrochloric acid) contained no oxygen.

Of course, supporters of the strong programme may seek to show that beliefs about reasons are themselves caused, in part, by social forces. Proponents of the program have included among "social forces" shared goals, interests, and research traditions. These social factors are present in virtue of a scientist's participation in groups and organizations. The first stage in a strong programme inquiry is to identify the relevant groups. However, the mere identification of a scientist's family relationships, nationality, class status, institutional affiliations, and research associates is insufficient. At least some of these factors must be shown to have contributed causally to that scientist's specific actions. Whether such factors can be shown to be "contributory causes" of individual evaluative decisions must be decided on a case-by-case basis. The burden of proof on supporters of the strong programme is heavy indeed.

Normative Naturalism

Normative naturalism is the position that evaluative standards and procedures arise within the practice of science, and are to be assessed in the same way that scientific theories are assessed—by reference to claims about the world. The normative naturalist views science and philosophy of science as a seamless whole. She is concerned to deny that philosophy of science is a "transcendent" discipline in which trans-historical, inviolable evaluative principles are superimposed upon the practice of science.

The *normative* naturalist maintains, nevertheless, that the standards developed within the philosophy of science have prescriptive status. Normative naturalism is a prescriptive enterprise whose acknowledged aim is to uncover standards for the appraisal of scientific theories and explanations. It is the normative naturalist position that such standards, like scientific theories themselves, have provisional status only. They are subject to correction or abandonment in the light of further experience.

Neurath's "Boat-Repair" Image

Otto Neurath suggested that the development of science resembles the continuing repair of a ship at sea.

> we are like sailors who have to rebuild their ship on the open sea, without ever being able to dismantle it in drydock and reconstruct it from the best components.[1]

Neurath's "boat-repair" image (the above is one of three such images he put forward[2]) illustrates the program of normative naturalism. Theories, evaluative standards and justificatory arguments on behalf of those standards all must be created, applied, and revised during the voyage. There is a continuing need for adjustment and revision of these aspects of scientific practice. In the absence of creative responses to the need for repairs, the ship may sink. The needed repairs must be carried out on the boat itself during the voyage. There is no transcendent standpoint (drydock) from which repairs may be made.

Neurath insisted that all claims about theories, evaluative standards, and justificatory arguments are subject to revision and replacement. No proposition within science is incorrigible. This includes statements that record observations.

Neurath referred to observation reports as "protocol sentences." He maintained that protocol sentences always include reference both to an observer (or recording instrument) and to an act of observation (recording). An example is "Al noted at 2.15 pm that Al observed at 2.14 pm that the Mercury meniscus in the glass tube before him was on line 8.2." There are various reasons why this protocol sentence may be rejected. Al may be observed to read the meniscus level from a sharp angle or in poor light. Al's prior protocol sentences may have been unreliable. Other viewers at the time may locate the meniscus level at 9.2. Al may have elected the value 8.2 in order to confirm a theory that he supports, and so forth. Neurath pointed out that the acceptance or rejection of a protocol sentence depends on judgments about the truth-status of other empirical hypotheses.

Neurath held that there is an analogy between the seaworthiness of the ship and coherence among protocol sentences, theories, evaluative standards, and justificatory arguments. Thus there is an inviolable principle that governs the repair process. At every point of the voyage, scientists must seek to resolve conflicts within their discipline. Protocol sentences, hypotheses, theories, and evaluative principles must be adjusted to form a coherent whole. Moreover, scientists need to respond to a perceived lack of coherence within the discipline, but also to social, economic, and political pressures from the larger community.

Suppose a scientist on the boat discovers a leak (lack of coherence). She recognizes that an evaluative decision is required. One obvious response is to apply an evaluative standard whose past applications have repaired leaks. Of course, the present evaluative situation is not precisely the same as it was in previous applications. The scientist may be justified in searching for a different evaluative standard.

Regardless of the evaluative standard selected, a supporting argument is required. Given the continuity displayed by the boat's voyage, a justificatory argument can be formulated in support of the standard that previously had been successful. viz.:

1. If the current evaluative situation resembles earlier evaluative situations, then one is justified in applying that evaluative standard that was effective in the earlier situations.

2. The present evaluative situation resembles those earlier evaluative situations.

Therefore, one is justified in applying the previously effective evaluative standard in the present evaluative situation.

Premise (1) is a general directive principle based on induction by simple enumeration. Unfortunately, there are important episodes in the history of science in which implementation of this directive principle failed to repair a leak. Examples include:

1. because the anomalous motion of Uranus was accounted for by postulating the existence of an additional planet (Neptune), so also account for the anomalous motion of Mercury by postulating the existence of another planet (Vulcan);
2. because numerous motions of bodies are accounted for by postulating central-force field interactions between bodies, so also account for electromagnetic induction by postulating central-force field interactions; and
3. because "complete explanation"—including both spatio-temporal description and a causal analysis—can be given for the motions of planets and billiard balls, so also a "complete explanation" should be given for an experimental result in the quantum domain. Quantum mechanics, on the Copenhagen interpretation, abandons this ideal of a "complete explanation." On the Copenhagen interpretation, the result of a measurement made in a particular experimental arrangement may be expressed in a spatio-temporal description or a causal analysis, but not both.[3]

One may defend the general inductive directive principle by maintaining that the "demands for repair" posed by the orbit of Mercury, electromagnetic induction, and quantum phenomena differ substantially from prior repair-demands. In each of these cases, there is a new type of repair-problem. To defend the general inductive principle in this way is to support a higher-level directive principle, viz.:

Apply a previously effective evaluative standard in all cases except those for which the evaluative situation is significantly different.

Whether a new evaluative situation is "significantly different" is assessed in the course of the ongoing repair process. The normative naturalist is comfortable at this point. The normative naturalist position is that new evaluative standards emerge, if they do emerge, in the context of new evaluative situations. Neurath insisted that all propositions within science, from protocol sentences to the most general evaluative principles, are subject to revision or rejection.

Normative-prescriptive decisions may be required at any point during the voyage. Normative status is conferred upon these decisions by the *inviolable* directive principle "seek coherence within the corpus of scientific propositions." Neurath's version of normative naturalism falls within the [N & I] box.

Quine's "Field-of-Force" Image

Neurath's version of normative naturalism received support from Willard van Orman Quine. Quine declared that

> we must not leap to the fatalistic conclusion that we are stuck with the conceptual framework we grew up in. We can change it bit by bit, plank by plank, though meanwhile there is nothing to carry us along but the evolving conceptual scheme itself. The philosopher's task was well compared by Neurath to that of a mariner who must rebuild his ship on the open sea.[1]

Quine superimposed a second image on the development of science. A high-level scientific theory is like a field of force that is subject to constraints imposed by the accumulation of experience. At the center of the force field are principles of extensive scope. At the periphery are specific claims about actual or possible observations. Quine noted that

> a recalcitrant experience can . . . be accommodated by any of various alternative reevaluations in various alternative quarters of the total system.[2]

Quine, like Neurath, took the demand for coherence to be the driving force behind scientific progress. If scientists perceive a failure of coherence they seek adjustments to restore it. The approach usually pursued is to change the force field at its periphery. This is a conservative strategy that seeks to restore coherence by making minimal changes to the theory. Occasionally, however, scientists reject or modify principles at the center of the force field. This took place, for instance, in the developments that led to the creation of relativity theory and quantum theory.

Quine insisted that the choice between tinkering at the periphery and changing principles at the center of the force field arises within the practice of science itself. There is no transcendent standpoint from which the proper evaluative response may be selected.

Kantorovich's Evolutionist Normative Naturalism

Aharon Kantorovich posed the following question about the philosophy of science

> Should [philosophy of science] adopt the task of *appraising* scientific claims: should it be content with the more modest aim of *describing* science, its methods and evolutionary patterns, or should it have both descriptive and predictive functions?[1]

Kantorovich chose the "descriptive-plus-predictive" alternative. But this is simply to select the normative-prescriptive position on the aim of the discipline. Of course a prescriptive philosophy of science will include a descriptive component. A second-order commentary requires a first-order subject matter. There can be a descriptivist philosophy of science that abjures normative-prescriptive pronouncements, but there cannot be a normative-prescriptive philosophy of science devoid of descriptions of scientific evaluative practice.

J. J. C. Smart suggested in 1972 that fictitious examples, rather than historical examples, could serve to illustrate the methodological principles of the philosophy of science.[2] However, even such extreme logicism would require reference to actual scientific practice to delimit the range of permissible hypothetical examples. Otherwise there would be no reason to accept the analysis as a "philosophy of science."

Kantorovich recommended a justificatory hierarchy with individual evaluative decisions at the base, evaluative standards at a higher level, and a "paradigm of rationality" at the apex. Justification is achieved by reaching "reflective equilibrium" between adjacent levels of the hierarchy.[3]

Kantorovich accepted the program of naturalism and sought to exclude *a priori* elements from the hierarchy. In particular, he emphasized that the paradigm of rationality, like evaluative standards and evaluative decisions, is subject to revision in the light of experience. The content of the paradigm is derived in part from developments within science and in part from general epistemological and metaphysical principles (which themselves are subject to change). Kantorovich examined three candidates—logicism, sociologism, and evolutionism. He selected evolutionary epistemology as the best qualified to serve as the source of foundational standards whose application justifies (in part) specific evaluative standards. An evaluative standard such as undesigned scope is justified provided that

1. it "complies with the paradigm of rationality,"[4] and
2. applications of the standard are "faithful to scientific practice."[5]

Local and Global Applications of Evaluative Standards

Richard Burian, Robert Richardson, and Wim van der Steen affirmed the position of normative naturalism in a paper on evaluative practice in genetics (1996). They applied Arthur Fine's distinction between "local" and "global" appeals to realism to questions about meaning and reference.

Fine had observed that within the context of a particular research program scientists regularly seek answers to questions about truth and existence, e.g., "Is it true that the offspring of blue-eyed parents invariably have blue eyes?" "Do there exist specific structures that are passed on from parents to offspring?" To pose such questions is to create conditions within which scientific progress may be achieved. This is the case even though standards for certifying truth and existence within a discipline may change over time.

By contrast, to debate whether "science-as-a-whole" is true or whether the "entities hypothesized by science" really exist is to engage in speculation tangential to the purposes of science. The global realist argues that the universe has a structure that is (largely) independent of being observed, and that the best scientific theories succeed in representing that structure. Science thus achieves truth, or at least converges upon truth, and the entities hypothesized by true theories do exist. The global antirealist denies these claims, insisting that the predictive success of science is insufficient to ground claims about a structural isomorphism of science and reality.

Fine recommended disengagement from global questions about truth and existence. It suffices for the interpretation of science that we adopt a "natural ontological attitude" (NOA) that accepts science "as it is." NOA accepts those scientific results that are certified by the internal standards of the discipline. The certification is always provisional, of course. In particular, the standards of appraisal within a scientific domain are subject to change. Nevertheless, the NOA position is that the achievements of science be taken to be on a par with the findings of common sense.[1]

Burian, Richardson, and van der Steen extended Fine's "global–local" distinction to questions about meaning and reference. Their conclusion was that

one cannot understand the development of science except in terms of the interaction of partially conflicting local theories.[2]

Inflexible commitment to a global theory of reference is an impediment to this endeavor to understand science. Burian et al. maintained, in particular, that it would be stultifying to apply universally either a strictly descriptivist theory of reference or a strictly causal theory of reference. Interpretations of science that rely on a descriptivist understanding of reference emphasize discontinuities within the development of science. Interpretations of science that rely on a causal account of reference, by contrast, emphasize continuity within the development of science.

Burian et al. suggested that the recent history of genetics provides support for the above conclusion. Geneticists have taken two approaches to the definition of "gene." The first approach is to define "gene" in functional terms. The second approach is to incorporate considerations of structure into the definition. Thus two directive principles have been influential within genetics:

R1. Genes can be defined in terms of function, indifferent to structure.
R2. Genes can be defined in terms of function and an associated structure.[3]

On the Burian-Richardson-van der Steen reconstruction of the recent history of genetics, the period before 1909 was marked by a failure to distinguish clearly the material basis of inheritance from the observed characteristics of offspring. In 1909 Johannsen achieved important clarification by drawing a distinction between genotype and phenotype.

Much research during the period 1909–1925 proceeded on the basis of a functional understanding of the genotype. Morgan, Muller, Beadle, and others achieved valuable results upon implementation of Rl. In the 1930s and 1940s, most geneticists adopted R2 as a directive principle. Genes initially were described incorrectly as proteins, and subsequently described as strands of nucleic acid.

Burian, Richardson, and van der Steen cautioned that it would be wrong to conclude that directive principle R2 has supplanted principle Rl within current research. A number of current formulations of the gene concept are sufficiently vague and ambiguous that the investigator may adopt either principle Rl or principle R2 depending on context. Considered from the functional standpoint (Rl) a gene is whatever "exhibits Mendelian inheritance and affects the phenotype uniformly."[4]

Considered from a structural-functional standpoint (R2) a gene is a sequence of nucleic acids which codes for RNA, or for proteins, or which controls transcription.[5] Given such ambiguity, one cannot assign a specific "conceptual content" and a specific "empirical content" to the concept.

If asked, in the abstract, whether the claim that genes are nucleic acid sequences is empirical or conceptual, or whether it is analytic or synthetic, the only sensible response would be to decline the question. It is empirical, and conceptual. It is neither analytic nor synthetic. Locally, however, in particular contexts and at particular times, with particular issues at stake, researchers did take the trouble to disentangle the conceptual and empirical, or aspects of the conceptual and the empirical.[6]

In addition to providing directives for research in genetics, principles Rl and R2 also serve as schemata for subsequent historical reconstruction. Burian, Richardson, and van der Steen noted that a discovery concerning the hereditary material which expands our knowledge concerning the structure or composition of genes will count as empirical according to R1, yet according to R2 it will count as a discovery that we have mischaracterized genes, or, perhaps, that some sorts of entities also perform functions that are performed by genes. Yet these are not disagreements over facts or over the historical record; they are differences over how to interpret that record and the actual (agreed upon) findings. Rl and R2 can yield equivalent descriptions covering the same empirical data and encompassing the same historical changes, though they will disagree over how to partition factual and conceptual components of those changes. In this respect, it is a matter of indifference whether the discovery that DNA rather than protein is the genetic material counts as a conceptual change or not. We can describe the associated changes as changes in concepts or in beliefs. Under Rl it becomes a change in belief, and under R2 a change in meaning.[7]

Burian, Richardson, and van der Steen maintained that there are alternative ways to partition the empirical–conceptual boundary. They held moreover, that "no reconstruction is privileged in a context-independent, global way."[8]

This is a claim that could be made within an exclusively descriptivist philosophy of science. But Burian, Richardson, and van der Steen superimposed a normative superstructure upon the descriptivist claim. They insisted that the choice between Rl and R2 in a particular context of research is not arbitrary. In some cases at least, there are features of the research situation that make one decision about the empirical–conceptual boundary superior to another. They hold that one task of philosophy of science is to prescribe correct evaluative practice, and that correct practice may not coincide with the actual decisions made by scientists. It remains to be seen how the philosopher of science is to achieve this normative-prescriptive standpoint without appeal to global, context-independent principles.

Clearly the situation in genetics is quite different from the situation in quantum mechanics. Within quantum theory (on the Copenhagen interpretation), it is features of the experimental situation that determine the interpretation to be applied to the system between measurements. There always is a "preferred interpretation" for a given experimental arrangement—either spatio-temporal description or causal interaction (but not both). "Preference" is determined by applications of Heisenberg's uncertainty principle, a principle held to be inviolable within the Copenhagen interpretation. There is no comparable inviolable principle accepted by geneticists, and principles Rl and R2 do not provide mutually exclusive, but complementary, interpretations.

Burian, Richardson, and van der Steen referred, not to global inviolable principles, but to "pragmatic" and "heuristic" considerations. What is unclear is how such considerations dictate normative conclusions.

Burian, Richardson, and van der Steen maintained that, in certain contexts, application of principle Rl would have been preferable evaluative practice despite the fact that researchers actually applied R2. For instance, if there are empirical questions extant about the structure of an entity supposedly responsible for manifestation of a certain trait, it would be

> unwise to beg the empirical questions by embedding presuppositions about the relation of structure and function into the concepts employed.[9]

Selection of R1, a purely functional approach, would be preferable evaluative practice.

The methodological rule implicit in such judgments is

> to reduce the conceptual burden of a concept when that will allow the empirical resolution of a seemingly conceptual dispute.[10]

Unfortunately, applications of this methodological rule also may have adverse consequences. By shifting the empirical–conceptual boundary in the direction of the empirical, one may decrease explanatory power. At times a contrary approach is preferable. Burian, Richardson, and van der Steen concede that

> at least sometimes, by imposing an additional conceptual load it is possible to sharpen issues by imposing greater constraints on the problem at hand. By increasing conceptual load and conceptual power, one removes certain empirical questions from the table and constrains the space of solutions to open questions. This may well turn out to be a virtue in particular cases.[11]

This vacillation—sometimes maximize empirical content, sometimes maximize conceptual content—may suffice for a descriptivist philosophy of science. But what is needed for a normative position is some procedure to ascertain under what conditions, and in what direction, to shift the empirical–conceptual boundary in specific cases.

Burian, Richardson, and van der Steen are precluded from specifying such a procedure because they hold that evaluative practice can be judged to be correct only with respect to the purpose of inquiry. One can say that if an increase of explanatory power is the purpose of inquiry, then one ought . . ., or if an increase of empirical content is the purpose of inquiry, then one ought . . .

The normative content of this position extends no further than the claim that one ought not accept any global inviolable principle.

Philip Kitcher on Achieving Cognitive Virtue

Philip Kitcher sought to stake out a middle ground for normative naturalism. This middle ground is located between the extremes of normative-descriptive logical reconstructionism and a descriptivism that "abandons or relativizes" normative values.[1] Kitcher maintained that an important goal of normative naturalism is to show how normative-prescriptive standards are created and justified within the history of science. It might seem that this is a futile understanding, given the following argument:

1. Normative naturalism cannot include trans-historical inviolable evaluative principles.
2. The normative status of a proposed evaluative standard can be warranted only by appeal to trans-historical inviolable principles.

Therefore, normative naturalism cannot warrant the normative status of proposed evaluative standards.

Kitcher proposed a burden-of-proof shift. The normative naturalist is not required to show that warranted evaluative standards can arise from analysis of scientific practice. Rather, the critic is challenged to prove that consensus about evaluative practice does not occur in the history of science. Kitcher declared that

a strategy for defending traditional naturalism should become evident. With respect to the historical and contemporary cases, the aim will be to show that the case for continued divergence and indefinite underdetermination has not been made out . . . All that traditional naturalism needs to show is that resolution is ultimately achieved, in favor either of one of the originally contending parties or of some emerging alternative that somehow combines their merits.[2]

One obvious candidate for convergence that results in consensus is the requirement that a scientific theory be linked, however indirectly, to observables by operational definitions. After decades of debates about an "aether" that carries electromagnetic waves, convergence was reached upon the operational requirement that P. W. Bridgman called the "epistemological lesson" implied by Einstein's theory of special relativity.[3] According to Bridgman, one lesson learned in early-twentieth-century science is that every (non-logical) concept not linked to measuring procedures should be excluded from physics.

The link to instrumental procedures may be indirect. The stress in a solid, for instance, is linked to instrumental procedures by reference to calculations made from surface measurements of strains on its surface. The velocity of a gas molecule in a container is linked to measurements of temperature and pressure in virtue of its participation in the root-mean-square velocity of all the molecules. The Ψ-function is linked, *via* calculations involving the square of its absolute magnitude, to measurements of scattering distributions and electron-transition frequencies.

A second candidate for convergence is the determination of the values of physical constants. Measurements of melting points, atomic weights, thermal conductivities, et al. have converged over time on specific values. The same is the case for "theoretical constants" such as Avogadro's number, Planck's constant, Boltzmann's constant, and the velocity of light.

In many cases, convergence is achieved on the value of a particular constant from diverse types of measurement. For example, determinations of the value of Avogadro's number from studies of radioactivity, Brownian motion, the viscosity of gases, and black-body radiation converge upon the value 6.02×10.[4]

The history of science also reveals that consensus has been reached on certain ideas. Examples of ideas judged reliable are "atoms exist," "oppositely charged bodies attract one another," and "unsupported bodies fall to the earth." It seems appropriate to speak of "approximate truth" in such cases.

Kitcher maintained that we may be able to *show* what counts as "approximate truth" even if there is no trans-historical inviolable standard of truth to which to appeal. We learn about "approximate truths" by tracing developments within science.

Kitcher held that science may be judged to have made progress toward truth if developments within the discipline display a unification that increases our understanding of causal relations among phenomena. Such progress represents the attainment of "significant truth."[5]

The evaluative standard "unification that confers understanding of causal relations" is not a trans-historical principle imposed on science from above. It has been developed from, and receives support from, the analysis of episodes within the history of science.

Insofar as there is consensus within science on this principle, it may be accorded normative status. And if this is the highest-level normative principle in science, then those evaluative standards whose applications promote unification of our causal knowledge also have normative-prescriptive status.

The procedure that leads to the achievement of "significant truth" is not subject to change within Kitcher's philosophy of science. This procedure is a convergence that leads to consensus. Kitcher invoked "convergence that leads to consensus" as a trans-historical inviolable principle that establishes significant truth. It would seem that Kitcher's normative naturalism falls within the [N & I] box after all.

Shapere's Nonpresuppositionist Philosophy of Science

Dudley Shapere sought to develop a "nonpresuppositionist" philosophy of science. He noted that evaluative standards are developed in the course of the practice of science. These standards have prescriptive force. They specify how scientific theories ought to be formulated and appraised. No such standard is inviolable, however. Standards are subject to revision and replacement within the history of science. Shapere suggested a view of science

> according to which that enterprise involves no unalterable assumptions whatever, whether in the form of substantive beliefs, methods, rules, or concepts.[1]

Shapere's proposal would place the philosophy of science in the [N & ~I] box.

In studies undertaken in the 1970s, Shapere sought to uncover methodological principles implicit in the development of scientific domains. The concept of a "domain" is a category for the interpretation of the history of science. A domain is a body of information that comprises a unified subject matter. In a domain, items of information are associated such that:

1. the association is based on some relationship among the items;
2. there is something problematic about the items so related;
3. scientists regard the problem to be important; and
4. science is "ready" to deal with the problem.[2]

One domain is the collection of chemical elements. Nineteenth-century scientists discovered that there is a periodic order among elements when they are arranged in order of increasing weight. Shapere pointed out that this order implements a "principle of compositional reasoning." He declared that to the extent that a domain D satisfies the following conditions or some subset thereof, it is reasonable to expect (or demand) that a discrete compositional theory be sought for D.[3]

The relevant conditions are that the domain is periodically ordered and that the members of the domain have values that are multiples of a fundamental value.

A second domain is the stars. Late nineteenth-century astronomers catalogued stars by reference to the light that they emit. They took spectral color to be a measure of age. Young stars are blue-white. Old stars are red. Stars of intermediate age display other colors. Shapere noted that this ordering exhibits a "principle of evolutionary reasoning." This principle stipulates that a correlation be sought between the properties of the objects of a domain and a temporal ordering of these objects.[4]

What is the status of the principles thus uncovered? I. B. Cohen cautioned that the twentieth-century methodologist who seeks to reconstruct past developments must take care not to read present concerns back into the past. What counts are the principles that overtly or covertly inform research, and not the principles that we believe should have guided research.[5]

Shapere agreed that anachronistic reconstructions of the past are to be avoided. However, he was not content to restrict domain studies to an uncovering of patterns of reasoning implicit in past practice. He declared that

the rationality involved in specific cases is often *generalizable* as principles applicable in many other cases.[6]

The principle of compositional reasoning, for example, specifies a pattern that fits such domains as chemical combination, the spectral classification of elements, and Mendelian studies of heredity.[7] Hence, when Shapere declared that, given a domain that satisfies certain conditions, it is reasonable to expect (or demand) that a theory of a particular type be sought for that domain, he meant in part that, in a high percentage of cases, a theory of that type can be formulated for domains that satisfy those conditions. Thus far, he has made only a descriptive claim to be tested by appeal to the history of science.

However, Shapere also recommended the principles of compositional reasoning and evolutionary reasoning as principles directive of further research. He held that if a particular type of theory can be formulated for a high percentage of domains that satisfy certain conditions, then, given a new domain that satisfies these conditions, scientists *ought* to formulate a theory of that type. This, of course, is to move from a generalization about scientific practice to a normative conclusion about what counts as "good science."

How can this transition be justified? Shapere failed to provide a justification in his 1970s essays. Subsequently he admitted that in these essays he had recommended patterns of reasoning "without any indication of their source or justification," and that a justification is needed.[8] Justification of patterns of reasoning—*qua* directive principles that ought guide research—is itself a normative undertaking. Whatever the ambiguities of his earlier theory about domains, his post-1980 project was to develop a normative philosophy of science within which evaluative principles are formulated *and justified.*

Shapere placed three requirements on normative philosophy of science. These requirements presumably reflect our intuitions about scientific progress.

The first requirement is that philosophy of science recognize that evaluative standards and procedures are subject to change. It is indisputable that evaluative standards and procedures have changed during the history of science. Shapere emphasized that when tensions arise between theories and evaluative standards, either may be modified or abandoned. He noted that success of quark theory had been accompanied by a de-emphasis on the requirement of observability. In other cases, an evaluative standard is more highly prized than a particular theory, in which case it is the theory that is changed.[9]

The first requirement rules out any philosophy of science that includes immutable evaluative criteria. Shapere criticized philosophers of science who subscribe to an "inviolability thesis," according to which there exist standards of rationality that must be accepted before analysis of the scientific enterprise can take place.[10] There are no immutable standards by which science may be appraised. Criteria of acceptability, rules of procedure, epistemological and metaphysical assumptions, all are subject to change. Moreover, what counts as "rational" is itself subject to historical development. No supra-historical standpoint is available.

The second requirement is that the philosophy of science accommodate the possibility that knowledge is achieved within science. Shapere held that theory-replacement often constitutes a gain in knowledge and that the philosophy of science ought account for this possibility. He declared that

> an adequate philosophy of science must show how it is possible that we *might* know, without guaranteeing that we *must* know.[11]

To this end, he suggested the following interpretation of "knowing." Given theory *T*, "*x* knows that *T*" if, and only if, "*x* believes that *T* and *T* has been applied successfully over a period of time and no one has a specific doubt about *T*."[12]

Certain theories are known to be false and yet have extensive applications. Examples include Newtonian mechanics and the ideal gas law. Two types of specific doubts may arise with respect to such theories. One type of doubt is that the descriptions and predictions drawn from the theory do not match reality "closely enough." The second type of doubt is that contrary-to-fact conditional claims implied by the theory are erroneous. In the case of the ideal gas law, such doubts refer to the law's "description" of what *would* be the behavior of dimensionless molecules subject to zero intermolecular forces. Presumably, such theories remain "true" on Shapere's view so long as they continue to be applied in the absence of specific doubts of either kind. And continued application of such theories indicates the absence of such doubts.

The third requirement is that philosophy of science "preserve the objectivity and rationality of science."[13] Shapere acknowledged that logical reconstructionists were correct to accept this requirement. What is needed, he claimed, is a philosophy of science that accepts the second requirement above but not the inviolability thesis.

What is involved in "preserving the objectivity and rationality of science"? A minimal interpretation is that the requirement of "preservation" merely excludes any account of science that denies the existence of standards of appraisal. For instance, if Feyerabend were to take seriously his own suggestion that "anything goes,"[14] then application of the third requirement would exclude this position from the class of acceptable philosophies of science. A stronger interpretation of "preservation" is that an acceptable philosophy of science *exhibit* the objectivity and rationality of a substantial portion of science.

Shapere characterized his own philosophy of science as a "nonpresuppositionist alternative."[15] Basic to this nonpresuppositionist position is the conviction that all evaluative standards, methodological rules, epistemological assumptions, and metaphysical commitments are subject to change within the history of science. He suggested that

> we carry the historically-minded philosopher's insight to its fullest conclusion and maintain that there is absolutely nothing sacred and inviolable in science—that *everything* about it is in principle subject to alteration.[16]

At any given time the philosophy of science will include evaluative principles. Shapere himself recommended the following principles:

1. Scientists ought "be open to the possibility of altering or rejecting any of the ideas with which [they] approach nature."[17]
2. "Attend only to doubts that are directed at specific beliefs."[18]
3. "Employ as background beliefs only those which have proved successful and free of specific doubt."[19]
4. "Aim at the internalization of scientific reasoning."[20]
5. Accept a succession of concepts as related by "rational descent" if there exist reasons for the transition from one concept to the next.[21]
6. Accept the continued successive use of one and the same term if there exist "chain-of-reasoning" connections between usages (e.g., continued use of the term "electron" in the theories of Stoney, Thomson, and Feynman is justified because there were "reasons at each stage for the dropping or modification of some things that were said about electrons and the introduction of others)."[22]
7. Scientists ought distinguish and keep separate the problem of observational support for beliefs and the problem of the relationship between beliefs and perception. Shapere provided support for this principle by showing that a distinction between these two problems was recognized explicitly in debates about the detection of the neutrino.[23]

Shapere insisted that the above principles are provisional and subject to modification or rejection in the light of further experience. Moreover, he held that, at a given time, different, possibly competing, evaluative standards justifiably may be applied in different contexts. He contrasted this "local presuppositionism" with a "global presuppositionism" that features inviolable evaluative standards.[24]

Whether Shapere's philosophy of science is labeled "nonpresuppositionist" or a "local presuppositionism," it is clear that it satisfies his first requirement for an acceptable philosophy of science. Nonpresuppositionist philosophy of science also satisfies the second requirement. Given Shapere's pragmatic concept of "knowing," a strong case can be made for the conclusion that knowledge has been achieved within science. There are numerous low-level laws—e.g., the laws of Ohm, Coulomb, and Balmer—that are applied under restricted conditions in the absence of specific doubts about their reliability. And if knowledge has been achieved, then the possibility that knowledge be achieved is realized.

Shapere's conviction that the history of science is progressive led him to affirm "consensus within science" as a mark of knowledge. It remains to be specified whose doubts count and what period of time is sufficient.

Larry Laudan complained that Shapere had compromised his "nonpresuppositionist" philosophy of science by affirming success as a transcendent criterion of theory-change.[25] Shapere replied that concepts of success themselves have evolved within the history of science, and that there is no "super-standard" of success by which we can decide which concept of success is correct.[26] But Shapere did take consensus to determine success. Of course today's consensus may not hold tomorrow. Nevertheless, the principle that it is consensus that determines success is not itself subject to change. It would appear that Shapere's philosophy of science is not presuppositionless after all. It includes at least one inviolable principle.

Shapere's third requirement is that the philosophy of science exhibit the "objectivity and rationality" of science. There are three ways in which the philosophy of science might achieve this:

1. select trans-historical inviolable standards of scientific progress and show that the history of science is consistent with application of these standards;
2. show that transitions from standard S_1 at t_1 to S_2 at t_2 are rational, despite the absence of trans-historical standards; and
3. simply trace the history of scientific evaluative practice.

The first approach is not available within a philosophy of science that denies the existence of trans-historical evaluative standards. The third approach abandons the philosophy of science in favor of the history of science. A narrative about scientific evaluative practice may exhibit the objectivity and rationality of science even if this objectivity and rationality cannot be demonstrated. However, this exhibition, if it is achieved, is achieved by historical reconstruction and not by philosophical analysis.

Shapere selected the second approach. He maintained that

> there is often a chain of developments connecting the two different sets of criteria, a chain through which a "rational evolution" can be traced between the two.[27]

What is needed is a shift of focus from a classificatory concept of justification to a comparative concept of justification. One obvious suggestion is:

> P^*—the replacement of evaluative standard S_1 by S_2 is justified if, and only if, S_2 is superior to S_1 on the standards of rationality accepted at t_2.

If the transition from S_1 to S_2 satisfies criterion P^*, then it is rational to apply S_2 and not S_1 at t_2.

However, it may be the case that S_1 is superior to S_2 on the standards of rationality accepted at t_1. Why base judgments of comparative justification on standards of rationality accepted at the later time?

The obvious answer is that we associate an increasing objectivity and rationality of science with its historical development. To base judgments of comparative justification on standards that have been superseded would force us to revise our understanding of the relationship between the concepts "rationality" and "history of science."

To accept evaluative standard P^* is to accept the higher-level principle P^{**}:

> P^{**}—justification procedures are to be based on currently accepted standards of rationality.

Is P^{**} a trans-historical evaluative principle inconsistent with a nonpresuppositionist philosophy of science? Shapere maintained that all evaluative procedures are subject to change. Hence he would not preclude the subsequent abandonment of P^{**} in favor of a principle that enshrines the standards of rationality of some particular epoch.

But then why accept P^{**} rather than some alternative principle? A plausible answer is that acceptance of P^{**} has led to progress in the history of science. Shapere has insisted, however, that there is no trans-historical standard of scientific progress. Concepts of progress themselves have evolved within the history of science. He stated that

> what counts as a legitimate successor [of a methodological rule] . . . is determined by the content of science at a given time, by its rules, methods, substantive beliefs, and their interplay—so that, with the further evolution of science, what counts as a "legitimate successor" may be different from what counted as such earlier.[28]

Shapere maintained that if "chain-of-reasoning connections" can be found between successive stages of evaluative practice, each stage is a "legitimate successor" of its predecessor. But he emphasized that what counts as a reason is subject to change, and with it the characteristics of the chain linking successive stages.[29] Suppose a group of scientists replaces standard S_2 with standard S_v and claims that they have done so because application of S_v justifies Immanuel Velikovsky's conclusions about the history of the solar system. These conclusions include the claim that a comet ejected from Jupiter caused a parting of the Red Sea and a temporary cessation of the earth's rotation. Other scientists criticize this replacement as irrational, on the grounds that evaluative standards have not previously been selected by appeal to considerations of that sort. The

Velikovsky-supporters reply that justification of Velikovsky's conclusions are a unique, newly relevant requirement for evaluative standards. Since no *a priori* distinction between reasons and non-reasons is permitted, the Velikovsky-supporters' proposal can be assessed only in retrospect. If scientists subsequently ignore the suggestion that standard S_1 be applied, then the rationality of replacing S_2 with S_1 may be judged not to have been a rational move.

To resolve disputes in this way is to appeal to the history of science to warrant methodological practice. Shapere maintained that the ultimate appeal in disputes about methodology is to that which works, and that that which works is established by scientific practice. To defend this position is to accept the following principle:

> $P***$—what counts as rational in science is determined exclusively by developments in the history of science.

Shapere's proposed nonpresuppositionist, but normative, philosophy of science is impaled on the horns of a dilemma. The first horn is that if the objectivity and rationality of science can be established only by justification of transitions between evaluative standards, then nonpresuppositionist philosophy of science cannot establish the objectivity and rationality of science. The second horn is that if the objectivity and rationality of science can be established only by tracing the historical development of scientific evaluative practice, then nonpresuppositionist philosophy of science is subsumed under the history of science.

The first horn appears to be true. Suppose that an evaluative standard is comparatively justified at a particular time. One might accept the directive principle that the standard be applied until it no longer receives comparative justification. Shapere insisted that higher-level directive principles of this type also are learned within history. As such, they too are subject to abandonment should experience so dictate. Moreover, criteria for determining whether or not "experience dictates abandonment" of a methodological principle is itself subject to modification or abandonment.

The still higher-level principle that methodological directives are subject to abandonment should experience so dictate is itself subject to abandonment should experience so dictate, and so on. Shapere has denied that immutable standards of rationality can be specified at any level of analysis. There is no standpoint from which to distinguish justified transitions between evaluative standards and the actual course of development of science.

By contrast, Imre Lakatos and Larry Laudan protect their philosophies of science from the slide into historical relativism by accepting immutable standards of rationality. Lakatos appealed to "incorporation-with-corroborated-excess-content" to justify both sequences of theories within scientific research programs and sequences of historiographical research programs.[30]

Laudan also outlined an evaluative procedure to comparatively justify a philosophy of science. Laudan's evaluative procedure begins with the selection of a set of historical episodes that are judged to be progressive. These "preferred cases" (which are subject to change) single out standard cases that provide a point of comparison for other judgments about scientific rationality. Competing philosophies of science then are gauged on their ability to reconstruct these standard cases.[31]

The second horn also appears to be true. Tracing the historical development of scientific evaluative practice is a task for the historian of science. If the objectivity and rationality of science were revealed in a particular history of science, this historical reconstruction would have no normative-prescriptive force. Shapere conceded that the history of science cannot justify scientific evaluative practice. He declared that

> the mere fact that a certain scientist or group of scientists in the past thought and worked in certain ways cannot be used as a base for generalizing to an interpretation of science in general: for those ways may well have been altered drastically later, or even abandoned, or relegated to being "external" to science.[32]

Nor can the fact that scientists today accept certain evaluative standards justify the continued use of these standards. Shapere's program succeeds only if one horn of the dilemma is false, or there is some third way to establish the objectivity and rationality of science. It would seen that the [N & ~I] box remains empty.

Shapere did not succeed in formulating a nonpresuppositionist, yet normative, philosophy of science that satisfies his own requirements for an acceptable philosophy of science.

Laudan's Three Versions of Normative Naturalism

The Problem-Solving Model

Larry Laudan was skeptical of attempts to read "convergence upon truth" from the historical record. He noted, in particular, that consensus does not establish convergence upon truth. The consensus of scientists has proved to be wrong on numerous occasions. There were times during which there was general agreement on the existence of phlogiston and the aether. The same is the case for the conservation of mass, the conservation of parity, and the requirement that a complete explanation of motion in a particular experimental arrangement include both a spatio-temporal description and a causal analysis. Laudan complained, moreover, that

> no one has been able to say what it would mean to be "closer to truth," let alone to offer criteria for determining how we could assess such proximity.[1]

Laudan interpreted science not as a search for truth, but as a problem-solving enterprise. He distinguished empirical problems from conceptual problems.

Empirical Problems

Empirical problems are questions about the structure and interrelation of objects within a scientific domain. Laudan labeled empirical problems "first-order problems." Of course, there are no purely empirical problems. Empirical problems invariably involve theoretical considerations. Laudan was aware of this:

> Why, then, call them "empirical" problems at all? I do so because...even granting that their formulation will be influenced by theoretical commitments, it is nevertheless the case that we *treat* empirical problems as if they were problems about the world.[2]

Examples of empirical problems include the structure of the DNA molecule, the reactions between chemical elements, and the relation between types of geological strata and fossil remains.

What counts as an empirical problem is determined by the concerns of scientists at the time. At various times there were empirical problems about phlogiston, caloric, and the aether, despite the fact that these posited "domain-objects" do not exist.

Progress is made within a scientific domain when unsolved problems are addressed successfully. The discovery of scandium, gallium, and germanium resolved problems about missing elements in Mendeleeff's periodic table of the chemical elements. Malpighi's discovery of capillaries that connect arteries and veins contributed to scientific progress by resolving a problem for Harvey's theory of circulation of the blood. And the discovery of fossil remains of an *Archaeopteryx* contributed to scientific progress by filling in a "missing link" in the evolutionary progression of life forms.

Problem-solutions vary in importance. Laudan pointed out that the resolution of an anomaly is a particularly important problem-solution. Anomalous problems are

> those empirical problems which a *particular* theory has not solved, but which one or more of its competitors have.[3]

Laudan maintained that

> whenever an empirical problem, p, has been solved by any theory, then p thereafter constitutes an anomaly for every theory in the relevant domain which does not also solve p.[4]

Consider the problem of stellar parallax, an anomaly for the heliocentric theory of Aristarchus. This anomalous problem was "solved" by

1. Ptolemy, by formulating a geocentric theory;
2. Copernicus, by insisting on the great distance between the sun and the stars;
3. Tycho Brahe, by presenting a modified geocentric theory; and
4. Friedrich Bessel, by measuring the extremely small angle of parallax (1837).

Of course, there were problems other than stellar parallax for the proponents of heliocentrism. The career of the parallax anomaly is just one aspect of the history of the heliocentrism–geocentrism controversy.

In Laudan's usage, anomalous problems are problems that a rival theory has solved. Thus the tellurium–iodine atomic-weight inversion was an unsolved problem for Mendeleeff, but not an anomalous problem. No other theory had solved this problem. Similarly, discrepancies in the motions of the moon and Saturn were unsolved problems for Newton, but not anomalies. Regardless of how the line is drawn between anomalous problems and merely unsolved problems, Laudan surely is correct to insist that

> one of the hallmarks of scientific progress is the transformation of anomalous and unsolved problems into solved ones.[5]

Conceptual Problems

Science also progresses by the solution of conceptual problems. According to Laudan,

> conceptual problems are higher order questions about the well-foundedness of the conceptual structures (e.g., theories) which have been devised to answer the first order questions.[6]

Conceptual problems arise when

1. competing theories are proposed;
2. there is tension between a theory and the basic methodological presuppositions of the domain; or
3. a theory is in conflict with the prevalent world-view.[7]

A conceptual problem of type (1) is the nineteenth-century conflict between the particle theory and the wave theory of light. The two theories were mutually inconsistent, and yet each theory achieved some successes. The particle theory accounted for partial reflection; the wave theory accounted for diffraction and polarization.

A conceptual problem of type (2) is the incongruity between Newton's interpreted axiom system for mechanics and the inductivist theory of procedure affirmed by most scientists at the time. Newton himself praised inductivist methodology.[8]

A conceptual problem of type (3) is the tension between mathematical models that "save the appearances" of planetary motions and the Aristotelian world-view that requires each planet to move within a spherical shell concentric to the earth. This tension led medieval theorists to distinguish the "mathematical truth" of Ptolemy's models from the "physical truth" of the Aristotelian world-view. Since these two types of "truth" were mutually inconsistent, there was cognitive dissonance at the heart of medieval astronomy.

Laudan maintained that problem-solving effectiveness determines the acceptability of a scientific theory. He declared that

the overall problem-solving effectiveness of a theory is determined by assessing the number and importance of the empirical problems which the theory solves and deducting therefrom the number and importance of the anomalies and conceptual problems which the theory generates.[9]

The Problem-Solving Model and Normative Naturalism

What is the relevance of the problem-solving model of scientific progress to the program of normative naturalism? Laudan noted that a critic might complain that the problem-solving model is purely descriptive, lacking normative force. His response was to concede that "the *specific* parameters which constitute rationality are time-and-culture-dependent," but to insist that the *general* nature of rationality is not time-and-culture-dependent.[10]

The transition from the problem-solving model as a contribution to descriptive philosophy of science to the problem-solving model that delivers normative judgments is based on two claims:

1. certain developments in the history of science were rational, and
2. "the test of any particular putative model of rational choice is whether it can explicate the rationality assumed to be inherent in these developments."[11]

Laudan noted that we use evaluative standards

> not to explain the obvious cases of normative evaluation . . . but rather to aid us in that larger set of cases, where our pre-analytic judgments are unclear.[12]

Laudan outlined a procedure for determining the rationality of methodological decisions. The first step is to select a group of the "scientific elite." These are men and women whose judgments about science are highly regarded by their peers. The members of this group are asked to specify a set of "standard-case" developments in the history of science that are indisputably progressive. Those developments that receive unanimous support from the elite are accepted as a test-basis for subsequent evaluative judgments. Given competing theories of evaluative practice, it is rational to select that theory whose applications reconstruct the greater number of the standard cases.

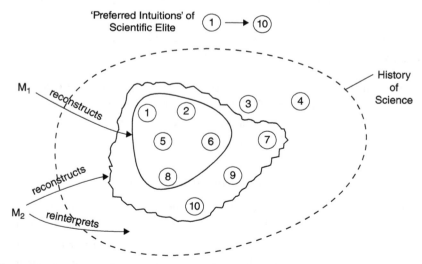

Figure 38 Laudan on the Evaluation of Competing Methodologies.

Source: John Losee, *A Historical Introduction to the Philosophy of Science* (Oxford: Oxford University Press, 2001), 239.

The choices of the scientific elite of 2100 may differ from the choices of today's elite. Standard-case developments are not set in stone. Nevertheless, at any given time Laudan's procedure provides a basis for the evaluation of competing methodologies.

A list of standard cases selected by today's elite might include the following:

1. heliocentrism is correct and geocentrism is incorrect on the evidence available in 1850;
2. Newtonian physics is superior to Aristotelian physics on the evidence available in 1750;
3. the Phlogiston theory was discredited by the evidence available in 1800;
4. the caloric theory of heat was discredited by the evidence available in 1900;
5. it was irrational to deny that atoms exist after 1920;
6. biparental heredity is correct, and ovism and animalculism are incorrect, on the evidence available in 1800;
7. it was rational to believe that fossils are the entombed remains of living organisms and are not the product of generative forces within the earth's surface by 1800: and
8. relativistic mechanics is superior to Newtonian mechanics on the evidence available in 1930.

Given a set of standard cases, one tests proposed rational reconstructions of science against the set. Laudan declared that

> the degree of adequacy of any theory of scientific appraisal is proportional to how many of the PI's (preferred intuitions) it can do justice to. The more of our deep intuitions a model of rationality can reconstruct, the more confident will we be that it is a sound explication of what we mean by "rationality."[13]

On this appraisal procedure, Galileo's directive principle—restrict the scope of physics to statements about primary qualities—is superior to Aristotle's directive principle that an acceptable scientific explanation of a process include a statement of its *telos*.

Laudan's justificatory procedure is not circular. But it does proceed along a spiral path. On his view, the philosophy of science and the history of science are interdependent disciplines. The history of science is the source of our intuitions about scientific growth, and the philosophy of science is a second-order commentary which sets forth the rational ideal embodied in these intuitions. The philosophy of science thus is dependent on the history of science for its subject matter. But, according to Laudan, the history of science also is dependent on the philosophy of science. The history of science is a reconstruction based on the rational ideal set forth in the philosophy of science.

The standard cases of scientific rationality are those cases that embody our "preferred intuitions." Laudan did not specify whose intuitions are to be satisfied. However, he did indicate that his methodology is *descriptive* with respect to the standard cases. Presumably, the "preferred intuitions" are those of the eminent scientists of the day. Laudan apparently believed that these scientists will agree on the selection of a set of standard cases. He compared judgments about scientific rationality to ethical judgments.

> As in ethics, so in philosophy of science: we invoke an elaborate set of norms, not to explain the obvious cases of normative evaluation (we do not need formal ethics to tell us whether the murder of a healthy child is moral), but rather to aid us in that larger set of cases where our pre-analytic judgments are unclear.[14]

Whether or not scientists agree about standard cases is an empirical question. If they do agree, then the consensus establishes the scientific rationality of the standard cases. Suppose scientists agree about standard cases, and agree also that one methodology provides the best reconstruction of these cases. Even so, the triumph of that methodology is provisional. The "preferred intuitions" of scientists may change, and new methodologies may be formulated.

Laudan conceded this. Nevertheless he maintained that the principles of the problem-solving model constitute the "general nature of rationality." The problem-solving model is not merely the best methodology currently available. Rather, this model

> transcends the particularities of the past by insisting that for all times and for all cultures, provided those cultures have a tradition of critical discussion (without which no culture can lay claim to rationality), rationality consists in accepting those research traditions which are the most effective problem solvers.[15]

It is Laudan's position that the criterion for rational choice among research traditions is the same at all times and in all cultures. This is the case despite the fact that specific standards of rationality—standards that determine what counts as a "problem" and a "problem-solution"—do depend on time and culture.

Laudan's distinction between the "general nature of rationality" and "specific standards of rationality" allows him to go between the horns of the normative-descriptive dilemma. The philosophy of science is normative insofar as it stipulates the criteria of rationality of research traditions; it is descriptive insofar as it focuses on theory-choices in terms of standards appropriate to the time in question.

Although the problem-solving model is the source of evaluative criteria, the model cannot be applied to yield an "instant assessment" of the rationality of theory-choice in individual cases. It is only the long-term fecundity of research traditions that is subject to evaluation for problem-solving effectiveness. Research traditions are

> general assumptions about the entities and processes in the domain of study, and about the appropriate methods to be used for investigating the problems and constructing the theories in that domain.[16]

Unfortunately, research traditions are not easy to evaluate. Laudan's "research traditions" resemble Kuhn's "disciplinary matrices" (the broad sense of "paradigm") and Lakatos' "research programs" in the following respects:

1. a currently successful research tradition subsequently may flounder; and
2. a currently ineffective research tradition may stage a comeback.

In the case of a currently ineffective research tradition, Feyerabend's observation is pertinent—"if you are permitted to wait, why not wait a little longer?"[17] What at first seems to be a dormant tradition, instead may be the beginning stage of a long-term progressive tradition.

This criticism was first directed at Lakatos. He replied that Feyerabend had conflated two distinct issues:

1. the methodological appraisal of a research program, and
2. the decision whether to continue to apply a research program.

Lakatos conceded that the appraisal-verdict on a research program may change over time. And he insisted that it is not the business of the philosopher of science to recommend research decisions to the scientist.[18]

Laudan would appear to be in the same situation. Despite the seeming finality of the claim that the problem-solving model stipulates the very "nature of rationality," he too conceded that the methodological appraisal of specific research traditions may change over time.

In addition, Laudan distinguished two modalities of appraisal: acceptance and pursuit.[19] In the modality of acceptance, it is rational to select the research tradition that has displayed the highest problem-solving effectiveness. In the modality of pursuit, it may be rational to select a less effective research tradition, provided that (1) its rate of progress is higher or (2) it is believed to have great problem-solving potential. Has Laudan conceded to Feyerabend that "anything goes"?[20] Not quite. Laudan did append a qualification. "Anything goes," provided only that it proves to be (in the long run) a contribution to the solution of some problem. It is doubtful, however, that this qualification removes the sting of Feyerabend's anarchism.

There are additional difficulties with Laudan's appraisal procedure. The composition of the selected elite may matter. Should it include eight physicists for every anthropologist, or eight anthropologists for every physicist? Are the standard-case developments equally important? Laudan stipulated only that one methodology is superior to a second methodology if it reconstructs a greater number of standard-case episodes. But suppose each methodology reconstructs five standard-case developments, but not the same five developments. No evaluative decision is forthcoming. In addition, there is the possibility that it is too easy for a methodology to reconstruct the standard-case episodes. Competing methodologies may reconstruct the same episodes and yet include different evaluative standards.

Laudan's commitment to the "standard-case" appraisal procedure was short-lived. In *Science and Values* (1984)[21] he abandoned that procedure in favor of a "reticulational model of justification."

The Reticulational Model

There are three levels of evaluation in the problem-solving model:

Level	Evaluation
3	Which research traditions account for the most standard-case episodes?
2	How is a conflict between a theory and the evaluative standards of a domain to be adjudicated?
1	Does theory T solve problem p?

Laudan's new position in *Science and Values* was to place theories, evaluative standards, and cognitive aims on the same level.

Laudan's Reticulational Model

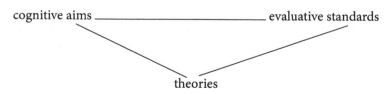

At each point in the development of a scientific domain an equilibrium must be established among these three elements. No one of the three elements has foundational status. Each element is subject to revision upon conflict with the others.

In many cases, theories are modified when they conflict with the evaluative standards current within a scientific domain, viz.

Standard	Original Theory	Revised Theory
Agreement with observations	Bohr theory of the hydrogen atom	Bohr-Sommerfeld theory
	Ideal gas law	Van der Waals' law
Deducibility of empirical laws	Descartes' vortex theory	Leibniz's harmonic vortex theory
Specify a causal mechanism for empirical laws	Kepler's three laws	Kepler's magnetic vortex theory
		Newton's gravitational attraction theory

In other cases, evaluative standards may be revised or abandoned in response to the success of a theory. The success achieved by the quantum theory in accounting for subatomic phenomena led Niels Bohr to challenge the prevailing standard for a complete explanation. The prevailing standard required that a complete explanation of an experimental study of motion specify both a spatio-temporal description and a causal analysis. Bohr proposed that the results of a particular experimental study of motion be subject to spatio-temporal description or causal analysis, but not both.

The history of science also reveals that tensions between theories and cognitive aims have been resolved, sometimes in favor of a theory and sometimes in favor of a cognitive aim. Laudan pointed out that the eighteenth-century cognitive aim that theories be restricted to claims about entities that can be observed was rejected over time because of the impact of theories that postulated the existence of unobservable entities. Laudan mentioned George LeSage's theory of gravitational corpuscles, David Hartley's theory of aethereal fluids in the nervous system, and Roger Boscovich's theory that forms of matter are the result of a distant-dependent force between point-particles.[22] He observed that, by the mid-nineteenth century,

> the "official" methodology of the scientific community had come to acknowledge the legitimacy of hypotheses about unobservable entities ... This

recognition becomes fully explicit in the writings of [John] Herschel and Whewell.[23]

Additional examples include:

1. The rejection of the LaPlacian cognitive aim of explaining electrical, magnetic, and chemical phenomena by positing distant-dependent central-force interactions between particles. This cognitive aim was abandoned after H. C. Oersted showed that the magnetic field around a current-carrying wire is circular at right angles to the wire (1829).
2. The rejection of the cognitive aim of restricting theories to those that display conservation of parity (left–right symmetry). This cognitive aim was enforced with success in the analysis of atomic spectra. It was abandoned, however, following studies of meson decay by C. N. Yang and T. D. Lee in 1956.

In other instances, theories that do not conform to cognitive aims are rejected or modified. A principal cognitive aim of the Cartesian methodology that was dominant in Europe during the seventeenth century was to explain motion by reference to impact or pressure. William Gilbert's theory that loadstones exert a non-corporeal magnetic force on one another and samples of iron was incompatible with this cognitive aim.[24] Descartes' response was to reject Gilbert's theory in favor of a theory about invisible screw-shaped particles that travel in a closed loop from one pole of a magnet to its opposite pole. When these screw-shaped particles pass through screw-shaped channels in an iron nail, they produce a lateral pressure that moves the nail toward the magnet.

Adherence to the cognitive aim of explaining motion as the result of action-by-contact also led Leibniz, de Molieres, and other scientists in the Cartesian tradition to insist that Newton's theory of gravitational attraction fails to provide a proper explanation of planetary motions. Newton's theory merely "saves the appearances" of planetary motions by "summarizing" Kepler's laws. Implementation of the Cartesian cognitive aim requires that there must be some aethereal fluid, or fluids, that press upon the planets. This pressure is the "true cause" of their motions.

Conflicts between cognitive aims and evaluative standards are infrequent within the history of science. However, they are important nevertheless.

There have been occasions when cognitive aims have been abandoned in favor of evaluative standards incompatible with them. Baconian inductivism was an important cognitive aim affirmed by seventeenth-century scientists.

Science should exhibit an inductive ascent from observation reports to laws to theories. The ascent is marked by increasing breadth of scope. Newton himself wrote that science should focus on relations among "manifest qualities" and theories induced from them.[25] Newton's praise of inductivism was at odds with the evaluative standard presupposed in his own *Principia*. This principle required only that experimental laws follow deductively from a theory, regardless of how the theory came to be formulated. The successes achieved by Newton's theory of mechanics led to the abandonment of inductivism as a required pattern for the discovery and expression of scientific knowledge.

There also have been occasions when cognitive aims have prevailed over evaluative standards. Albert Einstein's career reveals a preference for cognitive aims over evaluative standards in cases of conflict. Gerald Holton has emphasized that Einstein was committed to the cognitive ideals of simplicity, symmetry, and invariance. According to Holton, this commitment led Einstein to downplay Walter Kaufmann's experimental evidence that appeared to disconfirm the special theory of relativity.[26] Of course, Einstein did not deny the value of the evaluative standard "agreement with observations." He merely placed more weight of the cognitive aim of formulating theories that display symmetry and invariance. Some ten years later, Kaufmann's results were shown to be incorrect. Einstein's commitment to the above cognitive aim was reinforced.

On one level, Laudan's reticulational model is a descriptive philosophy of science. It provides a lens for viewing the history of science. Laudan maintained that the reticulational model is superior to Thomas Kuhn's "holistic" model in which theories, evaluative standards, and cognitive aims are replaced all at once. Laudan pointed out that if a Kuhnian "paradigm-shift"—in the broad sense of "disciplinary matrix" replacement—includes a wholesale replacement of theory(or theories), evaluative standard(s), and cognitive aim(s), then there is no objective basis for judging the rationality of the shift. Before the shift, scientists accepted [T, S, and A]. After the shift, they accepted [T*, S*, and A*]. Kuhn had conceded that there is no paradigm-neutral standpoint from which to judge a paradigm-shift. Laudan concluded that Kuhn's position is a retreat into evaluative relativism.

Gerald Doppelt complained that Laudan's reticulational model also is a version of evaluative relativism.[27] Laudan appeared to hold that any conflict resolution is acceptable provided that it restores equilibrium between apices of the evaluative triangle. For instance, it matters not whether theories are modified to accommodate evaluative standards, or *vice versa*.

Doppelt insisted that theories, evaluative standards, and cognitive aims are not of equal importance. Given a conflict between a theory and a cognitive aim, the usual approach should be to reject or modify the theory.

Laudan claimed normative-prescriptive status for the reticulational model. He provided a defense against the charge of evaluative relativism. He declared that the reticulational model

> is intended to be more than descriptive of existing practices. It purports to give a normatively viable characterization of how discussions about the nature of cognitive values should be conducted.[28]

To achieve normative content for the reticulational model, Laudan added two constraints on cognitive aims. These aims are required to be both consistent and realizable. Cognitive aims are subject to change, but consistency and realizability are inviolable constraints on the appropriate ways to resolve tensions within the evaluative triangle.

Unrealizable cognitive aims include:

1. Aristotle's requirement that the first principles of a science, and their deductive consequences, state necessary truths. Newton was correct to insist that science cannot establish that relations among phenomena cannot be other than they are, and
2. the aim of providing a deductive-nomological explanation for every relation among phenomena. Hempel was correct to insist that many scientific explanations fit an inductive-statistical pattern that is not reducible to a deductive-nomological form.

But even if realizability is accepted as a constraint, it is doubtful that the combination of consistency and realizability can confer normative-prescriptive status upon the reticulational model. The methodologist is directed to restore equilibrium in such a way that consistency and realizability are satisfied. However, tensions within the triad may be alleviated by modifying any one component, or combination of components. In many evaluative contexts there are numerous consistent and realizable adjustments that can be made to restore equilibrium. The methodologist needs to know, for a specific situation, whether to modify a theory, an evaluative standard, or a cognitive aim. The reticulational model merely restricts permissible modifications to those that preserve consistency and realizability. The seemingly intractable problem for normative naturalism is to justify the transition from naturalistic inquiry to standards whose implementation is required to achieve "good science."

Paul Thagard was optimistic on this issue. He declared that

> we can hope to generate from historical studies methodological principles whose sway can be extended normatively to cover the general practice of science.[29]

Thagard outlined a procedure to extract normative-prescriptive standards from an examination of case studies of scientific practice. He maintained that these standards can be justified by appeal to coherence. Given actual evaluative practices, the relevant background theories, and the goals of science, selection of a particular set of evaluative standards can be shown to produce a "maximally coherent" set of beliefs.

Thagard suggested "robustness," "accommodation," and "efficacy" as criteria of the comparative coherence of alternative sets of beliefs. "Robustness" is a measure of the range of evaluative practices accounted for by application of the set of evaluative standards. The greater the range of practices covered, the more robust the philosophy of science that contains the standards.

"Accommodation" is a measure of the extent to which appeal to background theories (psychological, sociological, et al.) explains those cases of evaluative practice that do not conform to applications of the evaluative standards. And "efficacy" is a measure of the degree to which application of the evaluative standards satisfies the cognitive aims of science.

Success at achieving "maximal coherence" requires the determination that a particular set of standards introduces more coherence within the total range of evaluative situations than does any other set. This is a difficult requirement to meet, particularly in virtue of the fact that evaluative standards often conflict. Agreement-with-observations and simplicity, for example, are evaluative standards widely believed to be applicable to scientific theories. But the application of these standards pulls in opposite directions. It is necessary to assign relative weights to these standards. In a particular evaluative situation, a 70–30 weighting of agreement-with-observations and simplicity may receive high marks on the three criteria of coherence. However, it would be hard to prove that no other weighting would be more effective. And even if it could be shown that maximal coherence has been achieved, one still might object that this does not constitute a warrant for crossing the "is/ought boundary." That a certain set of standards maximizes the coherence of beliefs with respect to a particular range of evaluative practice does not imply that those standards ought be prescribed to additional cases.

Means–End Relationships

Laudan presented a modification of the reticulational model in "Progress or Rationality? The Prospects for a Normative Naturalism" (1987).[30] He reaffirmed the position that "the aims of the scientific community change through time."[31] Neither cognitive aims nor evaluative standards are inviolable. Laudan now proposed that the two are related by means of hypothetical imperatives of the form

> if one seeks to realize cognitive aim *A*, then one ought to apply evaluative standard *S*.

Examples include:

Evaluative Standard	Hypothetical Imperative
Provide operational definitions of concepts	If the cognitive aim is to formulate theories from which predictions may be compared with observation reports, then require that the values of at least some of the terms of the theory be determined by results of instrumental operations.
Incorporation-with-corroborated-excess-content (Lakatos)	If the cognitive aim is that the transition from T_n to T_{n+1} within a research program be progressive, then require that: 1) T_{n+1} account for the prior successes of T_n; 2) T_{n+1} have greater empirical content than T_n; and 3) Some excess content of T_{n+1} be corroborated.
Problem-solving effectiveness (Laudan)	If progress depends on the solution of empirical and conceptual problems, then select from competing theories the one that solves the greatest number of (and/or most important) problems.
Undesigned scope	If the cognitive aim is to identify theories not subject to future rejection, then select those theories that receive unexpected support from quarters that initially seemed irrelevant or hostile to it.

If the cognitive aim expressed in the antecedent clause of one of these means–end relationships is taken to be true, then the consequent clause states an imperative to be realized. Whether or not acting on this imperative achieves the cognitive aim of the antecedent clause is an empirical question. Laudan stressed the importance of historical analysis. The philosopher of science needs to investigate whether acting on a proposed imperative has been more effective than the available alternatives in realizing the cognitive aim.

If it has been effective in the past, then it should be implemented in the present situation. Laudan recommended the following inductive directive principle:

> If actions of a particular sort, *m*, have consistently promoted certain cognitive ends, *e*, in the past, and rival actions, *n*, have failed to do so, then assume that future actions following the rule "if your aim is *e*, you ought to do *m*" are more likely to promote those ends than actions based on the rule "if your aim is *e*, you ought to do *n*."[32]

Robert Nola and Howard Sankey emphasized that

> there are no "ought" claims in the premises, but there are ought-claims in the principles arrived at in the conclusion. In this respect MIR [the meta-methodological inductive rule] is a level 3 metamethodological principle that sets out the conditions of warrant for P, a level 2 principle of method.[33]

Laudan took this inductive directive principle to be inviolable. He argued that this principle is itself a conclusion drawn from the study of the history of science. He declared that this rule

> is arguably assumed universally among philosophers of science, and thus has promise as a quasi-Archimedean standpoint.[34]

Individual standards and aims may change, but the inductive rule is taken to be inviolable. The inductive rule ensures normative status for Laudan's naturalism.

Appeal to the meta-metholological inductive rule blocks the potential regress of evaluative decisions, viz., application of standard *S* is warranted by appeal to rule *R* that stipulates how standards are to be evaluated; rule *R* is warranted by appeal to rule *R** that stipulates how rules that govern the selection of standards are to be selected; rule *R** is warranted by appeal to rule *R*** that stipulates how rules that stipulate how rules that govern the selection of standards are to be selected, etc.

Laudan stressed the normative-prescriptive aspect of his naturalism. He complained that

> Quine, Putnam, Hacking and Rorty, for different reasons, hold that the best we can do is to *describe* the methods used by natural scientists, since there is no room for a normative methodology which is prescriptive in character.[35]

It would appear that Laudan has reinstated a justificatory hierarchy incompatible with the reticulational model. The meta-methodological inductive rule occupies the apex of a hierarchy of evaluative principles.

It occupies this position in spite of the fact—recognized by Laudan—that some means–end correlations once held to be reliable have ceased to be so.

Examples include:

End	Means	Fail
Understand the motions of bodies	Postulate $1/R^n$ central forces	Electromagnetic induction
Predict the orbits of planets	In cases of discrepancy, postulate existence of an undiscovered planet (Neptune, Pluto)	Orbit of Mercury (Vulcan)
Understand processes in which products differ from reactants	Require conservation of mass in the process	Radioactive decay
Provide a complete explanation of an experimental result	Formulate both a spatio-temporal description and a causal analysis of the result	Quantum phenomena

It may objected that it is inappropriate to retain the general inductive directive principle in the face of such counterexamples. Some prior applications of the principle have led scientists astray. But then there are no guarantees available within science. One may have to live with a directive principle that is only "reliable on most occasions." Of course, after a previously reliable means–end correlation has been shown to have exceptions, then there no longer is a correlation. Subsequent applications of the inductive directive principle will not include that particular means–end correlation.

Of course, it is open to Laudan to take the inductive directive principle to be a contingent generalization subject to modification. The appropriate modification might be

P# = apply the general inductive principle except in those cases for which a formerly successful means–end correlation has ceased to be successful.

P# then might be selected as an inviolable principle, applications of which determine correct evaluative practice. However, *P#* would be of no use to the methodologist, since a current evaluative situation may, or not be, a case for which a formerly successful means–end correlation ceases to be successful.

Foundationalism

General Arguments against Foundationalism

Most philosophers of science have been dissatisfied with a purely descriptive approach to their discipline. They have preferred a normative-prescriptive alternative that issues recommendations about evaluative practice in science. Logical reconstructionism was a resolutely normative-prescriptive position.

Foundationalist philosophies of science formulated after its demise retained several emphases of the logical reconstructionist position.

Robert Audi presented an overview of the foundationalist position. On Audi's view, foundationalism is the position that the set of one's propositions (judgments, beliefs) includes a subset accepted without support from other propositions, and that non-foundational propositions receive support from the foundational subset.[1]

Candidates for foundational status may be divided into two classes:

1. sets of statements that provide support (deductive or inductive) for other statements within science (e.g., "primary experimental data," observation reports, empirical laws);
2. directive principles (evaluative standards, cognitive aims, rules of procedure).

A variety of positions can be developed within this framework, depending on the answers given to the following questions:

1. What is the content of the foundational propositions?
2. Are the foundational propositions held to be inviolable?
3. What counts as "support" within the set of propositions?

"Strong foundationalist" positions hold that a subset of propositions (or a directive principle) is inviolable. "Weak foundationalist" positions, by contrast, hold only that a subset of propositions (or a directive principle) is more reliable than the rest. Given prescriptive intent, strong foundationalist philosophies of science fall within the (N&I) box. Weak foundationalist philosophies of science, if there are such, would fall within the (N&—I) box.

The strongest relation of "support" provided by a set of foundational propositions would be deductive subsumption. In a formal system, a foundational subset (axioms) implies the non-foundational propositions (theorems) of the system.

But empirical science is not a formal system. Foundational propositions in science (if there are such) do not imply non-foundational propositions directly. Additional non-foundational premises are necessary—premises that provide information about the context of application. A proposition P qualifies as foundational provided that P, in conjunction with non-foundational propositions, is taken to determine the truth, high probability, or acceptability of additional non-foundational propositions. For example, suppose someone insisted that Archimedes' principle of buoyancy is a foundational principle that provides support for numerous non-foundational claims. The non-foundational proposition that "ice cubes float in water" does not follow deductively from the foundational principle alone. An additional premise about the relative densities of ice and water is required. A similar analysis is appropriate to applications of Fourier's theory of heat conduction. Fourier's theory stipulates the rate of heat transfer within an infinitely long slab of material of arbitrary conductivity, density, and specific heat (assuming that the finite length is "very long").

Thus if there are foundational propositions in science that provide deductive "support" for non-foundational propositions, they do so only on the assumption that other non-foundational propositions are accepted. It remains possible, of course, that there exist foundational propositions in science that provide inductive (rather than deductive) support for non-foundational propositions.

The word "foundation" suggests a Baconian view of science. Francis Bacon had maintained that the statements of science form a pyramid with observation reports at the base and the most inclusive generalizations at the apex. Bacon held that proper scientific method involves a stepwise inductive ascent from the base to the apex of the pyramid. He insisted that the collection of observations recorded in "natural histories" provides inductive support for the theoretical claims within science just as the foundation of a house provides material support for its upper stories. "Baconian foundationalism" is untenable. In the following discussion of "foundationalism," the term will be used in Audi's broad sense, such that the support provided by a subset of foundational statements may be either deductive or inductive.

A number of specific proposals for foundational status have been advanced. Before examining these proposals, it is appropriate to dismiss certain objections to the very idea of a foundational base for science.

Van Fraassen on the "Jesuit Argument"

Bas van Fraassen recently has called attention to arguments of the seventeenth-century Jesuit philosopher Francois Veron (as summarized by Richard Popkin and Paul Feyerabend). Veron's arguments were directed against the Protestant rule of faith—*sola scriptura*.[2] The position of *sola scriptura* is that passages of scripture alone determine

1. what can be known about the nature of man and his place in the universe;
2. proper conduct;
3. the resolution of theological disputes; and
4. the appropriate forms of worship.

Veron noted, first of all, that scripture itself cannot establish what writings are to be included as canonical. A selection of writings must be made and designated "foundational." But interpretation also is required. The canon of scripture comprises a sequential ordering of words. Scripture cannot itself establish the correct interpretation of this ordering.

Moreover, applications of the rule of faith require additional premises. "Do not remove the money from that safe" does not follow from "Thou shalt not steal" alone. It is necessary to establish that the removal would constitute a case of theft. Van Fraassen observed that in many instances, extrapolation is required. The interpreter of scripture needs to make inferences that go beyond the text itself. For example, the text does not state explicitly that Abraham has kidneys and elbows, but we are entitled to draw this inference upon appeal to our general background knowledge.

Veron concluded that Protestants could not implement successfully their "only scripture" principle. To use scripture as a foundational base, it is necessary to identify, interpret, and extrapolate. Veron urged the Catholic alternative. On this alternative, applications of scripture are guided by a continuous tradition of interpretation based on the authority of the Church, a tradition that derives from teachings of the Fathers and Doctors of the Church.

Van Fraassen adapted these arguments to refute a "classical empiricism" that takes as its rule of faith *sola experientia*. In van Fraassen's characterization,

> classical empiricism points to a unique putative source to ground and test all judgment: "experience is our sole source of information." A naive appeal to experience assumes that there is never any question about what the deliverances of experience actually are, nor about their meaning or significance. It assumes furthermore that the implications—namely, which theories are in accord with experience and which in conflict—is evident and unequivocal.[3]

Van Fraassen pointed out that to apply the rule *sola experientia* we must first identify those experiences that are veridical and then construct interpretations of them within our language. But to identify and interpret is to move beyond the foundational base provided by our direct awareness of the events that happen to us. He noted, moreover, that our interpretive language is "heavily laden with old beliefs and theories." Neither experience qua awareness of what happens nor experience qua judgments involved in this awareness can serve as the sole foundation for the justification of knowledge claims.

Van Fraassen's criticism is decisive against an "extreme foundationalism" that restricts all justification within a system to a comparison of knowledge claims to propositions deduced from the foundational subset alone. However, it is doubtful that anyone would wish to defend such a position. It is a well-known result achieved by Godel that there exist indemonstrable true statements within any axiom system of moderate complexity. Moreover, if it is evaluative standards, methodological rules, or cognitive aims that are selected as foundational, one cannot deduce statements about proper evaluative practice from these statements alone. Additional premises are required.

For example, it may be argued that theory T_1 is superior to theory T_2 on Lakatos' criterion "incorporation-with-corroborated-excess-content." An appropriate justification will include particular premises about the successful applications of each theory, and how the concepts "incorporation" and "excess content" apply in this specific evaluative context. Similar considerations apply to the justification of claims about "undesigned scope," statistical relevance, and the comparative entrenchment of predicates.

It is a big step from rejection of extreme foundationalist positions based on the rules *sola scriptura* or *sola experientia* to the rejection of foundationalism in general. And yet van Fraassen declared that

> we cannot be foundationalists, least of all in epistemology. We have to accept
> that, like Neurath's mariner at sea, we are historically situated, relying on our
> preunderstanding, our own language, and our prior opinions *as they are now*,
> and go on from there.[4]

However, Extreme foundationalism and Neurath's boat are not the only options for evaluative practice. A foundationalist need not claim that a subset of statements provides the sole justification for all non-foundational statements. Nor need the foundationalist maintain that foundational statements *alone* provide justification for non-foundational statements. The "Jesuit arguments" are not effective against a "moderate foundationalism" that holds only that there exist foundational principles that appear (together with context-specific information) in some justificatory arguments. Van Fraassen needs to discredit more moderate foundationalist positions before requiring that we board Neurath's boat. More comprehensive challenges to foundationalism may be developed from the standpoints of Quinean Holism, historical inquiry and normative analysis.

The Objection from Quinean Holism

Given a theoretical system, W. V. Quine argued that it always is possible to single out a subset of propositions that are held inviolable. Adjustments required during applications of the system then may be restricted to the other propositions within the system. A problem for foundationalism is that no subset of propositions is intrinsically foundational. Whatever foundational base is selected, alternatives can be formulated, provided only that no restrictions are placed on the complexity of adjustments allowed elsewhere in the system.

What holds of theoretical systems holds for philosophies of science as well. Laudan's reticulational model, minus the constraints of consistency and realizability, provides a good illustration. One may designate as "foundational" selected cognitive aims, and restrict adjustments to changes of methodological principles or theories. But one also may designate as "foundational" certain methodological principles, and restrict adjustments to changes of theories or cognitive aims.

Holism is a defensible position. The fact that it can be maintained consistently is a valuable corrective to the demand that we take certain principles to be inviolable because this is required by the structure of the universe. Nevertheless one may concede the holist's point but insist that considerations of complexity are important. There may be pragmatic reasons to prefer simple, straightforward adjustments to complex, convoluted adjustments. Quine himself endorsed this pragmatic gloss on holism. This concession to pragmatism removes the sting of the thesis that the selection of foundations is, *in principle*, arbitrary.

The Objection from Historical Considerations

It is clear from the historical record that some principles once held to be inviolable subsequently have been modified or abandoned. For example, methodologists no longer accept the LaPlacian directive principle "explain all mutual interactions by reference to forces acting along the line joining the centers of the bodies in question." Critics have taken historical evidence of this sort as undermining foundationalist philosophy of science. Of course, the foundationalist may reject the inductive leap from the abandonment of some methodological principles to the mutability of all such principles.

A stronger anti-foundationalist argument is that past commitments to supposedly inviolable principles have inhibited progress in science, and, for that reason, no such principles should be accorded inviolable status today. For this argument to be effective, the critic must show that the net impact of implementing a methodological principle was negative. Contrary-to-fact conditional judgments will be required. Consider the case of the Cartesian program to explain change exclusively in terms of contact-action. The critic needs to show both that the successes of the program were less important than its failures and that if the program had been abandoned earlier there would have been progress not in fact realized.

Clearly, much hinges on what is taken to count as "progress." There are various proposals available on this topic.[5] Whewell, Lakatos, and Laudan proposed measures of progress; Feyerabend denied that any such measures exist. In addition, historians offer divergent judgments about the importance of specific developments in science. Given these disagreements among interpreters of science, the general "inhibition-of-progress" charge is not an impediment to the development of specific foundationalist positions.

No one of these global challenges to foundationalism is decisive. However, the foundationalist needs to build a positive case for her *specific* recommendations about evaluative practice in science.

Gerald Doppelt on the Criterial Use of Evaluative Standards

In the course of a debate with Larry Laudan, Gerald Doppelt defended a foundationalist alternative. He claimed that there exist standards that stipulate the conditions of scientific rationality, standards that

> define the criteria of scientific truth, knowledge, evidence, proof, explanations, etc., for a certain tradition or community of scientific practice.[6]

Laudan had held that methodological rules are hypothetical imperatives that specify the best available ways to achieve acknowledged goals. Justification of such rules involves empirical considerations. Scientists do, and should, reject the claim that "*X* is more likely than its alternative to achieve *Y*," if *Z* produces *Y* more consistently. Laudan's instrumental conception of methodological rules is an expression of the naturalist program to unify evaluative practice in science and philosophy of science. Methodological rules, like scientific theories, are contingent generalizations subject to modification and abandonment in the light of further empirical evidence.

Doppelt maintained that the following directive principles have been accorded status as foundational standards at one time or another in the history of science:

1. prefer simple theories to complex ones;
2. accept a new theory only if it can explain all the successes of its predecessors;
3. reject inconsistent theories;
4. propose only falsifiable theories;
5. avoid theories that postulate unobservable entities;
6. prefer theories that make successful surprising predictions over theories which explain only what is already known; and
7. prefer theories that explain, or are confirmed by, a wide variety of phenomena distinct from those which they were initially introduced in order to explain.[7]

Doppelt did not claim that each of these methodological directives is justified. His concern, rather, was to establish that such directives function within science as more than predicates in goal-specific hypothetical imperatives. Whereas Laudan interpreted directive principle (6) instrumentally—viz., if you want theories likely to stand up successfully to subsequent testing, then accept only theories that have been used to make surprising confirmed predictions— Doppelt assigned foundational status to the principle, viz.:

only theories that make successful surprising predictions, in addition to explaining what is already known, can be reasonably regarded as empirically well-founded, genuinely explanatory of anything, true, and/or a real contribution to scientific knowledge.[8]

Laudan had conceded that after 150 years of debate about directive principle (6), there is no agreement among methodologists on its effectiveness in promoting the cognitive aim of reaching theories that are subsequently successful.[9]

Doppelt took the continuing debate over principle (6) to invalidate Laudan's instrumental view of methodological principles. Within a naturalistic methodology, rules are to be accepted or rejected on the basis of empirical evidence. But the empirical evidence in this case is inconclusive. Doppelt maintained that the reason for this is that the debate about the principle was, and is, a debate about the status of a *criterion* of scientific knowledge, and not a debate about a means–end connection. Those methodologists who rejected principle (6) did so because they accepted competing normative standards of truth and knowledge.

Doppelt maintained that directive principles, *qua* foundational standards, prescribe to, but are neither derived from nor justified by, the results of empirical inquiry. To apply a foundational standard is to provide support for specific evaluative decisions at the level of theory-choice. For instance, given directive principle (7), Maxwell's electromagnetic theory of light ought be chosen over Newton's corpuscular theory. And given directive principle (3), Newton's gravitational theory ought be chosen over Huygens' rotating-intersecting-vortices theory of terrestrial gravity.

Doppelt conceded that certain principles taken to be foundational at one time subsequently have been abandoned. An example is directive principle (5) above. Doppelt was concerned to establish, not that foundational principles are inviolable, but only that at any given time scientists embrace principles whose applications determine what counts as truth, knowledge, and evidence.

The "Strong Foundationalist" Position

Worrall on Foundational Methodological Principles

John Worrall maintained that a philosophy of science should include evaluative standards that are both foundational and inviolable. He agreed with Doppelt that evaluative standards have been, and ought to be, used as criteria of explanation, evidential support, theory-replacement, etc. He also conceded that Kuhn, Laudan, and Shapere are correct that many evaluative standards have changed within the history of science. He insisted, nevertheless, that this is insufficient to establish that *every* such standard is subject to change.

Worrall contrasted his methodological foundationalism with Laudan's naturalism. He declared that

> our basic disagreement is this. Laudan ... "can see no grounds for holding any particular methodological rule—and certainly none with much punch or specificity to it—to be in principle immune from revision as we learn more about how to conduct inquiry." Whereas it seems to me clear that in order to make sense of the claim that we "*learn* more" about how to conduct inquiry, some core evaluative principles must be taken as fixed.[1]

Worrall maintained that there do exist general evaluative principles that are, and ought to be, in force at all stages of the development of science.

Candidates for Foundational Status

Intra-theoretical Consistency as Foundational

There is one evaluative standard the foundational status of which is non-controversial. A scientific explanation must have consistent premises. Any theory used in a proposed explanation must be internally consistent as well.

The principle of intra-theoretical consistency satisfies Audi's requirements for foundational status. It is accepted independently of support from other propositions in science, and it provides support for each theory that qualifies as logically consistent. "Support" in such instances consists of passing a test necessary for inclusion within science.

It is important to distinguish intra-theoretical consistency from inter-theoretical consistency. Scientists do not require that a new theory be consistent with established theories in order to be acceptable. However, they do require that a theory be internally consistent. Intra-theoretical consistency is a necessary condition for cognitive significance. If a theory includes inconsistent premises, then it implies every statement whatever. But a theory that implies both S and *not-S* provides support for neither. Since a theory with inconsistent premises implies everything, it cannot contribute to scientific understanding.

Imre Lakatos might seem to disagree. He suggested that "some of the greatest scientific research programmes have progressed on inconsistent foundations."[1]

The Bohr research program for interpreting the hydrogen atom is a case in point. Bohr used classical electromagnetic theory to calculate energy values for electron orbits, but denied that orbiting electrons radiate energy, as required by classical theory. Bohr's assumptions are not consistent with classical electromagnetic theory. Nevertheless, the specific Bohr theory of 1913 (a theory within his research program) is consistent, since it restricts energy release to transitions between discrete energy levels. Lakatos is correct that some research programs have involved inconsistent assumptions, but *no specific scientific theory* is acceptable if it incorporates inconsistent premises. Lakatos was well aware of this.

Since consistency is a necessary condition of explanation in general, philosophers of science implicitly accept its foundational status. What remain controversial are additional proposed foundations for science.

Observational Evidence as Foundational

It has become common within recent philosophy of science to concede that observation reports invariably are infected with theoretical considerations.

Thomas Kuhn, for instance, pointed out that Kepler and Tycho Brahe would give different, theory-based, descriptions of sunrise. And Feyerabend, Achinstein, and Quine have argued that the observational–theoretical distinction can be drawn in various ways, depending on the purpose at hand.[2]

Nevertheless, science is an empirical discipline. It would violate the spirit of the undertaking to systematically ignore the results of observation. In some respect these results are privileged. Scientists routinely adjudicate theoretical claims by reference to what is observed. Statements about what is observed are accepted not independently of all theoretical considerations, but independently of the truth-status of the theoretical claim under investigation. Consider scientists' responses to the following questions:

1. Is that solution acidic?
2. Is that substance stearic acid?
3. Is there more argon than neon in that gas mixture?

Affirmative answers are given provided that

1. a piece of paper containing blue litmus dye is placed in the solution and observed to turn red,
2. the substance is observed to melt at 69.4°c, and
3. a mass spectrograph is observed to display a higher peak at mass 40 than at mass 20. Defense of the above claims about chemical composition requires acceptance of "direct observation reports" about a color change, a change of state, and a trace on a piece of graph paper.

Let's return to Tycho and Kepler on the hilltop at dawn. Tycho's claim that "the sun rises from beneath the earth" and Kepler's claim that "the earth spins beneath the sun" are theory-laden claims. But there is a more basic observational level. Tycho and Kepler can readily agree that the distance between an orange disk and a green line increases over time. Perhaps "direct observation reports" of this type provide a foundational basis for the evaluation of other claims within science.

"Direct observation reports" record what is "immediately observed," often the results of instrumental operations—the positions of pointers on scales, the coincidence of lines, color changes, and so forth. These reports may be taken to be foundational in two ways. In the first place, "direct observation reports" provide support—*via* operational definitions—for claims about values assigned to scientific concepts. In the second place, direct observation reports are endpoints in a process of confirmation that provides evidential support for scientific laws and theories.

Operational Definitions

The conjunction of "direct observation reports" and a suitable operational definition implies a value of the operationally defined concept, viz.:

$$\frac{Oa\&Ra}{(x)\ [Ox \supset (Cx \equiv Rx)]}$$

$$\therefore Ca$$

where Ox = is a case in which operations O are performed,

Rx = is a case in which results R occur, and

Cx = is a case in which concept C applies.

Operational definitions have been stipulated for dispositional concepts like "soluble," "pressure," and "electric field strength." In the case of "soluble" the relevant operation is the placing of a substance in a liquid, and the relevant result is the dissolving of that substance. The "direct observation reports" of these events are a foundational base for the claims that "substance S is soluble."

Since a great number of scientific concepts are linked to statements of direct observation reports by operational definitions, it is natural to inquire whether every scientific concept ought be operationally defined. If this strong operationalist thesis were correct it would provide very strong support for foundationalism. The truth-status of every claim about a *bona fide* scientific concept would be determined by the truth or falsity of "direct observation reports." The language of science would be subdivided into foundational sentences whose truth or falsity is determined directly by observation and sentences whose truth or falsity is determined by logical relations to the foundational base.

The strong operationalist thesis is untenable, however. Although the truth or falsity of statements that assign values to contextually defined concepts is determined by data that record operations performed and results achieved, the truth or falsity of statements about certain theoretical concepts is not so determined. There exist *bona fide* scientific concepts not correlated *via* operational definitions to statements about "direct observation reports" (see p. 8).

This is fatal to a "strong operationalist thesis." However, it would be fatal to foundationalism itself only if foundationalism requires that *every* non-foundational claim receive support from the foundational set. A weak "observation-report foundationalism" remains a defensible position. "Direct observation reports" are accepted (provisionally) independently of other claims within science, and *at least some* of these non-foundational claims receive support from the set of "direct observation reports." A particular direct observation report may be dismissed from the foundational set, but there must be *some* basic observation reports that play a foundational role. This is a trans-historical inviolable principle, a necessary condition of the possibility of empirical science.

Van Fraassen's "Constructive Empiricism"

Bas van Fraassen's "constructive empiricism" is an influential version of observation-report foundationalism. Van Fraassen took statements about "observables"—concepts whose values can be determined by the unaided human senses—to constitute the foundational test-basis for the remainder of the statements within science. According to van Fraassen, theories that refer to "non-observables" specify models from which a set of "empirical substructures" may be delineated. It is these empirical substructures that are compared with the foundational reports about observations and experiments. If there is agreement, the theory in question is said to be "empirically adequate."[3]

Consider Mendel's theory of heredity. The theory makes claims about "characters" (genes) that do not qualify as observables on van Fraassen's criterion. However, the empirical adequacy of the theory may be gauged by comparing its empirical substructures (e.g. empirically interpreted Punnett squares) with observed phenotype ratios.

Van Fraassen maintained that empirical adequacy should be *the* cognitive aim of theory-construction. He did not deny that theoretical claims about non-observables may be true, but he recommended an agnostic position on this issue. For the purposes of science, it suffices to establish the empirical adequacy of a theory.

Foundationalism and Confirmation

A second way in which observational evidence may be taken to be foundational is as "end-points" of a confirmation process that provides support for scientific laws and theories. A number of philosophers of science have found this to be an attractive position.

Moritz Schlick defended a strong foundationalist position on confirmation. He maintained that the deductive testing of theories terminates in "confirmations," foundational experiences that are accepted without support from other beliefs.[4] A confirmation is a "judgment" of the form "here, now, so and so" which expresses the sense of a present gesture. A confirmation is an experience and not a statement that can be true or false. When the demonstratives "here" and "now" are written down, the resulting statement does not correctly represent the confirmation-experience. Schlick held nevertheless that confirmation-experiences *qua* experiences are not subject to doubt, and that they provide a secure foundational base for the testing of theories.[5]

Karl Popper agreed with Schlick's emphasis on deductive testing, but disagreed with Schlick's identification of its foundational base. According to Popper, the appropriate base is not confirmation-experiences but "basic statements." A basic statement stipulates that an intersubjectively observable event has occurred in a specific region of space and time. Basic statements record what is "directly observed" within a specific region at a particular time—the positions of pointers on a scale, the coincidence of lines, color changes, and the like. Examples include "that object now is green," "the liquid there now is cloudy," and "the top of that Mercury column now is on 6.7." For such statements, it is assumed that the relevant observation and recording can be carried out in such a way that further analysis is unnecessary. This is not to claim that basic statements invariably are true. Such factors as parallax, static, and human error may be present. But if specified recording procedures are followed, basic statements normally are accepted without reference to further observational procedures.[6]

In this respect, the role of basic statements in science is analogous to that of fingerprints in a criminal investigation. The detective in charge of a murder investigation may support his conclusion that the butler is guilty by stating that the fingerprints on the murder weapon match those of the butler. If the inspector were asked how he knows that the prints match, he might point to various loops and swirls which are similar in the two sets of prints. If he then were asked how he knows that the reference set of prints is really that of the butler, he might take another impression of the butler's fingers in the presence of the skeptic. It would be a radical skeptic indeed who viewed the taking of the impression and then asked, "but how do you know that these are really the butler's prints?" The analysis of the truth-status of a statement in terms of an appeal to other statements does not go on indefinitely. Eventually a point is reached at which the truth of a statement is accepted without a demand for further analysis. This is not to say that it is impossible to push the analysis further. We might check to make sure that the butler was not wearing skin-tight gloves on which prints were engraved, and so on.

Popper emphasized that the acceptance of basic statements is a matter of convention. We decide in a particular case not to press for further analysis.[7] Thus the foundation provided for science by its test-basis is "weak."

Popper insisted that the support provided by basic statements occurs in the context of the deductive testing of theories. Science does not proceed by induction from basic statements to generalizations.

At one time, many philosophers of science believed that a theory-independent set of observation statements could be identified and utilized as a test-basis for theoretical claims. However, considerations introduced by Feyerabend, Achinstein, Quine, and others have shown that this position is untenable.[8] The "observational–theoretical" distinction is context-dependent. Nevertheless a weak version of "observation-statement foundationalism" may be defensible. The weak version is that for a given test situation there often is a set of observation statements that is theory-neutral with respect to the theory under test. For example,

1. melting-point determinations—dependent on thermodynamic theory—provide a test-basis for the identification of organic compounds;
2. spectrographic findings—dependent on optical theory—provide a test-basis for theories about the chemical composition of the atmospheres of stars; and
3. chemical analyses of geological strata provide a test-basis for theories about the origin of the strata.

No claim need be made that the observational evidence in such cases is independent of all theoretical considerations. Nor is it necessary to claim that *every* bit of observational evidence accepted as providing support for a theory is free of infection from that very theory.

In Popper's version of theory-testing, the methodologist is to seek confrontations that maximize the likelihood of falsification. This requires exposure to the most severe tests that can be devised. Moreover, the methodologist is required to reject conventionalist strategies that would protect a theory from falsification.

Implementation of the falsificationist methodology is a process that takes place over time. Whether, and at what point, a practitioner has ceased to conform to its requirements is subject to debate. Methodological falsificationism requires that a scientist continually seek to test her theories. However, it does not require that she abandon a theory upon receipt of one particular negative test result. Popper conceded that the confrontation with negative evidence is never so compelling that one must regard the relevant theory to be false.[9]

There are various responses to *prima facie* disconfirming evidence that are consistent with the *long-range* commitment to expose a theory to the possibility of falsification. The history of science reveals that these responses are sometimes fruitful and sometimes not fruitful.

Scientists sometimes ignore evidence that appears to disconfirm a theory. Mendeleeff ignored evidence that the atomic weight of tellurium is greater than the atomic weight of iodine.[10] Einstein ignored Kaufmann's "experimental disconfirmation" of the special theory of relativity.[11] Subsequent developments proved that Mendeleeff and Einstein made the right decisions.

Scientists sometimes modify a theory to accommodate evidence that appears to count against it. Leverrier posited the existence of a trans-Uranic planet to accommodate the evidence that the orbit of Uranus failed to conform to the predictions of Newtonian gravitation theory.[12] Pauli and Fermi posited the existence of a new type of particle—the neutrino—to accommodate the apparent disconfirmation of conservation of energy in β-decay.[13] Subsequent developments justified these decisions.

On the other hand, Clairaut's revision of Newton's law of gravitational attraction to read "$F = k_1 \, m_1 \, m_2/r^2 + k_2 \, m_1 \, m_2/r^4$" in order to accommodate discrepancies between Newtonian calculations and the observed motion of the moon proved not to be fruitful.[14] The same was the case for Leverrier's suggestion that there is a planet (Vulcan) between Mercury and the sun that accounts for observed discrepancies between calculations from Newtonian theory and the motion of Mercury.[15]

David Miller maintained that methodological falsificationism is an effective procedure to eliminate false conjectures (over time). He denied, however, that such elimination justifies those conjectures that remain. The status of conjectures that thus far have survived destruction-tests is "not yet falsified" rather than "proved to be true." Miller recommended an anti-justificationist version of falsificationism. He declared that

> Falsificationism proposes a simple general method for undertaking the task of uncovering the universe's uniformities. But unlike all traditional and almost all modem forms of inductivism, it is not justificationist. It pursues not certain truth or justified truth, but plain unvarnished truth.[16]

He invited critics of methodological falsificationism to show that its recommendations—exclude non-falsifiable conjectures, devise and apply severe tests, avoid conventionalist strategies—are ineffective in the search for what is true about the world.[17]

The justificationist will object to this burden-of-proof shift. What is important, from the justificationist standpoint, is to establish that certain truth-claims are justified. It is not enough to show that a claim *may be* true (viz., that the claim has not yet been shown to be false).

Nelson Goodman's work on confirmation (see pp. 123–126) established that the set of observation reports that participate in confirmation relations is restricted to statements about "entrenched predicates."[18] Historical inquiry is required to certify the content of the set of "useful" observation reports. "Grue-type" predicates are to be excluded upon appeal to their lack of a track record of prior usage.

Goodman's work does not invalidate observation-report foundationalism. Observation reports still may be accepted without appeal to higher-level claims. And higher-level claims still may receive support from observation reports. What Goodman has shown is that the relation of evidential support is more complex than had been believed.

Thus far, two inviolable principles have been identified as constituents of a strong foundationalist position. The first inviolable principle is that scientific theories must be internally consistent. The second inviolable principle is that some terms in a scientific theory must be linked—*via* operational definitions—to statements that record primary experimental data. There must exist direct observation reports that are accepted without support from other propositions in science. These direct observation reports support—*via* confirmation relations—other propositions in science.

Empirical Laws as Foundational

Feigl on the Stability of Empirical Laws

Ernest Nagel noted that empirical laws possess a stability that scientific theories lack. Empirically confirmed laws often survive the demise of the successive theories that are invented to account for them.[1] To do justice to this distinction, Herbert Feigl suggested that the test-basis for scientific theories be located in low-level empirical laws rather than observation reports. Feigl's candidates for inclusion in the test-basis are from physics and chemistry. He cited laws of chemical composition, refraction, radioactive decay, and thermal conductivity, as well as the laws of Ohm, Ampere, Biot-Savard, Faraday, Kirchhoff, and Balmer.[2]

Feigl conceded that the laws he had listed presuppose antecedently accepted conceptual frameworks. Moreover, many empirical laws presuppose theories about the operation of scientific instruments. Thus empirical laws are not formulated independently of all theoretical considerations.

Nevertheless, empirical laws often are free of infection from the high-level theories they support. Empirical laws are accepted independently of high-level theories, and these laws do provide support for selected theories. (For illustrations from the history of science, see p. 157.)

Of course, empirical laws do not themselves have foundational status. They are not accepted independently of support from other propositions within science. Indeed, it is values recorded in observation reports that lead scientists to confer lawful status on certain correlations.

Individual empirical laws are subject to revision and even abandonment with the accumulation of experience. But the set of empirical laws accepted within a domain of science at a particular time may be treated as foundational by accepting the following principle as inviolable:

> If there exists a body of empirical laws in a domain of science, then evaluate theories in that domain by reference to those laws.

An augmented "empirical-law foundationalism" that includes this directive principle has considerable normative-prescriptive force for those domains of science that do contain well-established laws. Scientists are directed to take empirical laws into account in assessing the merits of theories. It would be methodologically improper to ignore well-established laws. However, what counts as the "evaluation of theories by reference to laws" remains to be specified.

It is clear that theories are underdetermined by empirical laws. Wilfrid Sellars pointed out in 1961 that what theories explain is why physical systems obey particular empirical laws to the extent that they do.[3] The kinetic theory of gases, for instance, explains why real gas samples obey the laws of Boyle, Charles, etc., to the extent that they do. On this view, the "evaluation of a theory by reference to empirical laws" involves a contrary-to-fact conditional claim—"if a gas *were* composed of elastically colliding point-particles subject to no forces other than Newtonian momentum transfer, then the ideal gas law would be obeyed." Physicists hold that the kinetic theory explains why gases obey the ideal gas law (to the extent that they do) despite the facts that molecules are not point-particles and that intermolecular forces are not zero.

Scientific theories stipulate the properties of *idealized* systems. Ronald Giere emphasized that the state of an idealized system is specified by an assignment of values to an appropriate set of variables.[4] For an ideal gas the appropriate state-variables are pressure, volume, and temperature; for Newtonian dynamics the appropriate state-variables are mass, position, velocity, acceleration, and force. A scientific theory postulates relations among its state-variables. These relations often include reference to the temporal unfolding of the system.

According to Giere, an idealized system achieves explanatory significance when combined with a "theoretical hypothesis" that certain physical systems exhibit the structure of the idealized system. Of course, there always is a degree of approximation involved. Whether the degree of approximation achieved is sufficiently high to confer explanatory value upon a theory cannot be resolved by appeal to some inviolable standard. The answer given by scientists in a particular case depends, in part, on the availability of viable competing theories.

Cartwright on Causal Patterns

Nancy Cartwright agreed with Feigl that well-confirmed empirical laws have privileged status within science. She maintained that

> phenomenological laws are meant to describe, and they often succeed reasonably well.[5]

Cartwright noted that theories, by contrast, lack descriptive adequacy. Indeed, she maintained that the explanatory power of high-level theories invariably is achieved at the expense of descriptive adequacy.

Of course, the fundamental laws of a high-level theory are true of its model-objects, but these laws include implicit *ceteris paribus* clauses. The law of universal gravitational attraction, for instance, describes motions only if no forces other than gravitational forces are present. Cartwright noted that

> no charged objects will behave just as the law of universal gravitation says; and any massive object will constitute a counterexample to Coulomb's law.[6]

The truth of the matter is that the fundamental laws of physics lie;

> rendered as descriptions of facts, they are false, amended to be true, they lose their fundamental explanatory force.[7]

Nevertheless it might seem that phenomenological laws provide differential inductive support for competing theories. Given a body of empirical laws and alternative theories that purport to explain why these laws hold, "inference to the best explanation" may single out that theory which is closest to the truth.

Cartwright was skeptical. She insisted that there is no one-to-one correspondence between explanatory success and truth. Indeed the situation is worse. Cartwright built a case for the position that truth and explanation are antithetical epistemological virtues within science.

Cartwright maintained that, despite this tension, empirical laws sometimes provide good grounds for claims about the existence of the causal entities postulated by a theory. Consider, for example, the phenomenological laws that correlate the shapes of curves observed in a Wilson Cloud Chamber with the strength of the applied magnetic field. These laws constitute evidence for theoretical causal laws that attribute the curves to the passage through the chamber of particles of specified mass, charge, and velocity. Cartwright held that it is appropriate to attribute a causal role to these theoretical entities because they have been applied successfully in other experimental studies. She cited with approval Ian Hacking's position. According to Hacking, if scientists successfully make use of the causal powers of postulated entities in the investigation of other entities or processes, then it is reasonable to accept the postulated entities as real. Paraphrasing Hacking, "if you can spray them, then they are real."[8] For instance, electrons that pass through a cloud chamber are "non-observables" in van Fraassen's sense, but since they are manipulated to investigate other structures (e.g., in electron microscopes), the claim that "electrons exist" is true.

Thus, whereas van Fraassen restricted truth-claims to statements about observables, maintaining an agnostic stance on claims about theoretical entities, Cartwright allowed the predication of "truth" to a certain type of theoretical claim. In certain cases it is true that there exist theoretical entities that are causally responsible for the observed regularities recorded in empirical laws. Cartwright emphasized that the search for such causal patterns is a primary aim of science. In support of this position, Rom Harre declared that

> successful and unsuccessful attempts at the manipulation of unobserved entities and structures is as important a feature of the epistemology of science as are successful and unsuccessful attempts to perceive the processes that produce observable phenomena.[9]

The empirical-law foundationalism of Cartwright, Hacking, and Harre is consistent with the position of entity realism.[10]

Foundationalism and Evolutionary Biology

An augmented empirical-law foundationalism is more convincing for physics and chemistry than for biology. There are thousands of low-level laws of chemical composition, thermodynamic equilibria and electromagnetic interaction. Some theoretical claims derive support from these laws. However, there is no comparable set of well-established empirical laws within evolutionary biology. If there were such laws, they would state the differential results of the relative adaptedness of organisms within environments of specified characteristics.

Robert Brandon has emphasized that the structure of Darwinian evolutionary theory "does not fit any existing philosophical paradigm for scientific theories."[11] According to Brandon, the basis of evolutionary theory is not a set of foundational empirical laws, but rather the schematic principle

> *a* is better adapted than *b* in E, if, and only if, *a* is better able to survive and reproduce in E than is *b*.[12]

Brandon interpreted this schematic principle as stipulating a propensity interpretation of "relative adaptedness."

Brandon conceded that the schematic principle itself lacks specific biological content. He noted, however, that it becomes applicable to evolutionary contexts when

1. there exist biological entities that are "chance set-ups with respect to reproduction";[13]
2. these entities differ with respect to adaptedness in a common selective environment; and
3. the differences in adaptedness are heritable.

If these three conditions are realized, then application of the schematic principle are empirically significant.

Brandon's principle of "relative adaptedness" functions as a directive principle in the science of biology. In this respect, it plays a role similar to Newton's second axiom of motion in physics. The physicist is directed to (1) identify a case of acceleration, and (2) discover the force, or combination of forces, responsible for the acceleration (given that "F = m a"). Similarly, the biologist is directed to (1) identify an adaptive scenario for which the three conditions are realized, and (2) apply the principle of relative adaptedness to this scenario.

The unavailability of a test-basis of empirical laws within evolutionary biology is not fatal to a foundationalist interpretation of the discipline. There remain the foundational principles that prescribe intra-theoretical consistency and operational links to primary experimental data, as well as the directive principle of relative adaptedness that is implicit in the theory of natural selection.

Evaluative Standards as Foundational

Kuhn's Set of Evaluative Standards

One evaluative standard is universally acknowledged to have foundational status. This standard is intra-theoretical consistency. Are there others? Thomas Kuhn noted that, in addition to consistency, scientists have appraised theories by reference to the following standards:

1. agreement with observations,
2. simplicity,
3. breadth of scope,
4. conceptual integration, and
5. fertility.[1]

It would be easy to produce examples in which each of these standards has been applied. However, to assign foundational status to these standards on this basis would be premature. Applications of these standards raise significant problems.

Kuhn emphasized that these standards are vague and often conflict. Considered as a set, they do not constitute a foundational base for scientific evaluative practice. Kuhn declared that this set of standards is "an insufficient basis for a shared algorithm of choice."[2] It remains possible, however, that one or more individual standard does qualify as foundational.

Agreement-with-observations and simplicity are antithetical requirements. Agreement with observations is maximized only at the expense of simplicity, and *vice versa*. Which of the two standards is selected to prevail in a given evaluative context depends on the cognitive aim selected. Neither individual standard, nor their conjunction, is foundational.

Scientists often view favorably an increase of scope achieved by successive theories. For instance, Arnold Sommerfeld received credit for expanding the scope of the Bohr theory of the hydrogen atom to include electron transitions to and from elliptical orbits.

Unfortunately, breadth of scope can be increased in less satisfactory ways. One theory may have greater scope than a second theory simply as a result of conjunction. The conjunction of plate tectonics theory and the kinetic theory of gases has greater scope than either conjunct, but conjoining the two theories does not increase the acceptability of either theory. Breadth of scope is a useful evaluative standard only if irrelevant expansions of a theory's domain are excluded. Since the relevance of an increase of scope depends on further analysis, it is not a foundational evaluative standard.

Conceptual integration is achieved whenever (1) "mere facts" are converted into "facts required by theory" or (2) breadth of scope is increased in a manner that promotes unification. An example of the first type is the replacement of an earth-centered planetary system. Copernicus placed the sun at the center of the system of planets, which system includes the earth. By so doing, he converted facts about the frequency and extent of retrograde motions to facts required by his theory. Copernicus' theory required that retrograde motion occurs more frequently for Jupiter than Mars, and that the extent of this motion is greater for Mars than Jupiter.

Examples of the second type are Newton's explanation of both terrestrial motions and celestial motions by reference to a principle of *universal* gravitational attraction, and Maxwell's explanation of both electrical phenomena and magnetic phenomena by reference to a general electromagnetic theory.

These examples of conceptual integration are impressive. However, N. R. Campbell has shown how facts arbitrarily may be converted into facts required by theory. As noted above (pp. 52–53), Campbell invented premises that imply the law of electrical conductivity in metals.[3] Surely Campbell's theory does not count as an achievement of conceptual integration. Since conceptual integration can be established only by excluding "explanations" of this kind, it is not a principle "accepted without support from other propositions." Conceptual integration does not fulfill Audi's requirements for foundational status.

The evaluative standard "fertility" is satisfied when a theory is extended to new types of applications over time. Newton's mechanics, quantum theory, and the theory of organic evolution have proved fertile in this sense. It remains to be specified, however, how "newness" is to be determined. Something more is required than additional confirming instances of the same type that the theory initially was proposed to explain.

William Whewell suggested (1847) that

> the evidence in favour of our induction is of a much higher and more forcible character when it enables us to explain and determine cases of a kind different from those which were contemplated in the formation of our hypothesis.[4]

Whewell spoke of such cases as instances of the display of "undesigned scope." Undesigned scope involves both novelty—the "eureka" effect—and extension to a new range of application.

His principal example was LaPlace's extension of the theory of heat transfer to account for the discrepancy between calculated and observed values of the velocity of sound. LaPlace's extension of the theory of heat transfer to the propagation of sound was unexpected, but one could argue that it was not an extension of the scope of the theory. The propagation of sound is just another case of the compression and expansion of an elastic fluid.

Whewell invoked historical considerations in support of undesigned scope. He wrote that if a theory

> of itself and without adjustment for the purpose gives the rule and reason of a class of facts not contemplated in the construction, we have a criterion of its reality, which has never yet produced in favour of a falsehood.[5]

Whewell's position is that no theory ever has been discarded after having displayed undesigned scope. This historical fact, if it is a fact, supposedly provides support for the evaluative standard. To take seriously this argument would be to base the status of undesigned scope on other propositions in science, and hence to deny foundational status to this evaluative standard. However, one may exclude Whewell's history-of-science defense and affirm the undesigned-scope standard independently of support from other propositions in science. It then would qualify as a foundational standard.

An additional evaluative standard, championed by Jean Perrin (1913), is the convergence of different types of experimental determinations on a specific result. Perrin called attention to various attempts to measure Avogadro's constant N.[6] N is the number of molecules in a gram-molecular weight of an element or compound. The term N occurs in the equations of theories of Brownian motion, electrolytic deposition, black-body radiation, radioactive decay, the viscosity of gases, and other theories. These are theories about diverse physical and chemical processes. Experimental data calculated from these theories converge on the value 6.02×10^{23} molecules/gram-molecular weight. Perrin argued that this agreement on the value of N provides important support for the hypothesis that matter is composed of molecules.

Max Planck subsequently (1920) emphasized the convergence of various experimental determinations of the value of h, the constant that bears his name.[7] Calculations from the equations of absorption spectra and emission spectra, black-body radiation, the photoelectric effect, and other processes agree on the value 6.62×10^{-27} erg – sec. This agreement on the value of h provides support for the general hypothesis that energy is quantized.

In addition, diverse techniques for measuring molecular weights often converge on a specific value. These techniques include vapor density measurements, static light scattering, mass spectrometry, freezing-point depression, and boiling-point elevation.

Undesigned scope and convergence of different types of experimental determinations are evaluative standards that may be affirmed independently of support from other propositions in science. They qualify as foundational principles on that score. However, applications of these standards depend on decisions about differences of type among experimental procedures. Unfortunately, these standards are only applicable to a small number of evaluative situations in science.

Lakatos on an Inviolable Evaluative Standard for Theory-Replacement

Thomas Kuhn's *The Structure of Scientific Revolutions* (1962, 1970) called attention to the following questions:

1. Under what conditions is theory-replacement rational?
2. What makes a sequence of scientific developments progressive?

In essays published in the 1960s and 1970s, Imre Lakatos developed answers to these questions.[8]

Lakatos held that the philosophy of science is a normative discipline, one task of which is to specify evaluative standards for theory-replacement. He noted that application of such criteria to developments in the history of science generates a reconstruction of scientific progress. In this reconstruction, instances of rational theory-replacement ("internal history of science") are distinguished from other instances of theory-replacement ("external history of science").

Lakatos insisted that normative appraisals be directed not at individual theories, but rather, upon "scientific research programs." A scientific research program is developed as a sequence of theories that implement a set of basic assumptions. The basic assumptions include a central core of axioms and principles, which are held to be inviolable throughout the development of the program. Successive theories adapt the core to empirical findings through the selection and application of auxiliary hypotheses.

An example cited by Lakatos is the Newtonian research program for astronomy. The core principles are the three axioms of motion and the law of universal gravitational attraction. Newton and his followers appended a succession of auxiliary hypotheses to the core in the process of seeking agreement with data on planetary motions. Among these hypotheses are (1) an asymmetric mass distribution of the Earth, (2) a theory of perturbations to approximate three-body interactions, and (3) the existence of a trans-Uranic planet.

Lakatos recommended an evaluative standard to assess sequences of theories within a scientific research program. This standard is "incorporation-with-corroborated-excess-content." It stipulates that the replacement of T_{n-1} with T is justified provided that

1. T_n accounts for the previous success of T_{n-1};
2. T_n has greater empirical content than T_{n-1}; and
3. Some of the excess content of T_n has been corroborated.[9]

Transitions that qualify as "progressive" on this criterion include

Ideal gas law	→	Van der Waals' equation
Bohr's theory of the hydrogen atom	→	Bohr-Summerfeld theory
Mendeleeff's theory (periodicity	→	Moseley's theory (periodicity
based on atomic weights)	→	based on atomic numbers)

In each case, predecessor and successor are related such that incorporation-with-corroborated-excess-content is achieved.

Lakatos suggested that applications of the incorporation criterion be restricted to the sequence of theories generated by a research program. He insisted that the refusal to accept apparently negative evidence as counting against the core principles of a research program often *promotes* progress. For example, astronomers were correct to pursue the Newtonian program in spite of the initial failure of calculations to reproduce the observed orbit of Uranus. So long as modifications of the auxiliary hypotheses generate theories that incorporate their predecessors and achieve additional success, the research program is progressive.[10]

Paul Feyerabend objected that important cases of theory-replacement within the history of science display explanatory overlap rather than incorporation.[11] He noted, for example, that the relationship between Newtonian gravitation theory and Cartesian vortex theory is not one of incorporation. The vortex theory explains the unidirectionality of planetary revolutions. Newtonian gravitational theory does not. Hence there is explanatory loss as well as explanatory gain in the transition from vortex theory to gravitation theory. Nevertheless the transition was a progressive development.

Feyerabend was correct that transitions from Cartesian vortex theory to Newtonian gravitation theory, and from classical physics to quantum physics and relativity theory, do not satisfy Lakatos' incorporation criterion. But these transitions are from one research program to another. Lakatos put forward the incorporation criterion as a criterion of the comparative acceptability of theories *within* a research program. Feyerbend's criticism is off-target. Nevertheless, he has shown that the incorporation criterion is not a necessary condition of theory-replacement *in general*.

Lakatos' incorporation criterion is not a necessary condition of theory-replacement within programs either. Progress often is made within a research program by improving the fit between empirical data and predictions drawn from theory without increasing the range of phenomena covered. And sometimes progressive theory-replacement involves a reduction of scope. An example is a research program to account for the properties of solutions of electrolytes. The replacement of Arrhenius' theory (1887) with the theories of Debye-Huckel (1923) and Onsager (1926) involves a narrowing of scope from electrolytes in general to strong electrolytes.[12]

Satisfaction of the incorporation criterion clearly is not necessary for progress within a scientific research program. But there remains the possibility that satisfaction of the criterion is a sufficient condition of progressive theory-replacement. Lakatos' criterion may yet be an inviolable evaluative standard. A foundationalist might claim that the incorporation criterion singles out instances of comparative acceptability within scientific research programs, just as *modus ponens* singles out valid instances of deductive reasoning. If T_2 subsumes every empirical law subsumed by T_1, and an additional law as well, or if T_2 explains everything explained by T_1, and additional phenomena as well, then T_2 is more acceptable than T_1.

However, there are both logical considerations and historical evidence against taking the incorporation criterion to be a sufficient condition of theory-replacement. The logical point is that if X is a well-confirmed theory consistent with the central core of the research program, whose last theory is T_1, then "$T_2 = (T_1 \& X)$" is more acceptable than T_1 alone on the incorporation criterion. But *ad hoc* conjunctions of this sort do not contribute to progress in science.

A historical counter-case to the sufficient-condition thesis is the Proutian research program discussed by Lakatos himself.[13] Prout's program was to show that the atomic weights of the chemical elements are exact multiples of the atomic weight of hydrogen. Prout accepted Newton's theory that all matter is inertially homogeneous. On this theory, it follows that, if the hydrogen atom is the minimum unit of matter, then the carbon atom contains 12 units and the oxygen atom contains 16 units.

Initially, Prout's program seemed promising. The atomic weights of nitrogen (14) and sulfur (32) provided support for the program. However, the atomic weight of chlorine (35.5) was not an integral multiple of the atomic weight of hydrogen. To accommodate this discrepancy, J. B. A. Dumas suggested that the basic building block of the elements has a mass equal to one-half the mass of hydrogen.[14] Dumas' modification accounted for all the successes achieved earlier and for the atomic weight of chlorine as well.

As more accurate data became available, it became obvious that a building block of one-half the atomic weight of hydrogen will not do. Dumas elected to continue the Proutian program by assigning to the basic building block of the elements an atomic weight one-fourth the atomic weight of hydrogen. Adding this auxiliary hypothesis within the Proutian program made possible the accommodation of the atomic weights of additional elements. The modifications made by Dumas qualifies as achieving "incorporation-with-corroborated-excess-content" within a specific research program. However, these developments clearly were not progressive. Long before we reach a "basic building block" of one-hundredth the atomic weight of hydrogen, Prout's initially bold scientific research program no longer is of interest.

"Incorporation-with-corroborated-excess-content" is neither a necessary condition nor a sufficient condition of acceptability for theory-replacement within a scientific research program. It may be accorded foundational status nevertheless. The incorporation standard may be affirmed independently of support from other propositions in science, and it may be cited in support of transitions between successive theories in scientific research programs. This is the case in spite of the fact that there are occasional non-progressive instances of incorporation-with-corroborated-excess-content within such programs.

Cognitive Aims as Foundational

McMullin on Justification and Cognitive Aims

Ernan McMullin developed a strong foundationalist position that assigns inviolable status to the basic cognitive aims of science. These aims are "predictive accuracy" and "explanatory power." McMullin declared that these cognitive aims serve to define the activity of science itself, in part at least. He placed these inviolable cognitive aims on the top rung of a justificatory ladder[1] (see Figure 39).

Specific evaluative standards are located on the second rung of the ladder. McMullin cited "consistency," "conceptual integration (coherence)," "unifying power," and "fertility." Presumably, the second rung also includes such standards as "undesigned scope" (Herschel), "consilience" (Whewell), and "problem-solving effectiveness" (Laudan).

McMullin assigned instrumental value to evaluative standards. The selection and application of a standard is a means to achieve the cognitive aims of science. To justify an evaluative standard is to show that its application promotes realization of predictive accuracy or explanatory power.

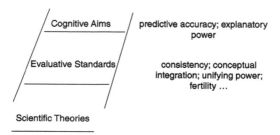

Figure 39 McMullin's Justificatory Ladder.

McMullin noted that there are two ways to justify evaluative standards. The first way is to stipulate and argue for conditions necessary and sufficient for a "good scientific theory." On this approach, a particular evaluative standard is to be justified by reference to a general philosophical theory about prediction and explanation.

The second way is to look at the historical record. Applications of a standard may be shown to have been associated with cases of theory-replacement that increased predictive accuracy or explanatory power. If this is the case, then continued use of the standard is taken to be justified. Standards such as conceptual integration, fertility, and unification presumably receive justification in this way.

McMullin declared that "ideally, both ways need to be followed, each serving as a check for the other."[2] The justification of evaluative standards is provisional, of course. Past success does not guarantee future success. Conditions may be about to change. McMullin conceded that the Humean position is correct. But he insisted that "demonstration is not what is called for" in the philosophy of science.[3]

McMullin's version of the justificatory ladder has great appeal. Most scientists and interpreters of science would agree that predictive accuracy and explanatory power are fundamental goals of science. Why not take these goals to be fixed points in the justificatory process? Specific evaluative standards then could be assessed as means to the realization of the goals.

There is a difficulty, however. The goals are quite vague. Consider "explanatory power." Minimally, to accept this goal is to affirm that scientists ought provide answers to "why-questions." However, this does not take us very far. Shapere was correct to insist that what kinds of why-questions are important is itself learned in the course of the development of science. For instance, we have learned to pose questions about undesigned scope, statistical-relevance relations, and unification. We believe that these factors, and others, are relevant to the "explanatory power" of theories.

But if the content of "explanatory power" and "predictive accuracy" is supplied by the history of evaluative practice in science then the cognitive aims of science are not fixed points in the justificatory process. Rather, these aims, like evaluative standards and theories, are subject to change within the development of science. McMullin's position then would be indistinguishable from that of Shapere. What is needed, for a foundationalist account, is an assignment of content to "predictive accuracy" and "explanatory power"—an assignment that holds independently of developments in science. The case is not altogether hopeless. Whatever the specific content of "explanatory power," every acceptable explanation has logically consistent premises. And whatever the specific content of "predictive accuracy," evidence that e_1 is not followed by an f provides no support for the rule "predict that if an event of type e occurs, an event of type f will follow." But considerations of this type provide only very weak constraints on these cognitive aims.

A second difficulty for McMullin's proposal is that the aims "predictive accuracy" and "explanatory power" do not always coincide. What counts in a particular evaluative context is not simply that predictive accuracy and explanatory power are acknowledged to be aims of science. Rather, it is the relative emphasis placed on each aim that is most important in the evaluative situation.

There have been evaluative episodes within the history of science in which predictive accuracy is emphasized almost to the exclusion of explanatory power. The tradition of "saving the appearances" in mathematical astronomy is a case in point. Within the tradition that derives from Ptolemy's *Almagest*, the prediction of planetary positions along the zodiac was taken to be the principal aim of astronomy. Andreas Osiander's statement of this aim is well known. According to Osiander, the astronomer

> must conceive and devise, since he cannot in any way attain to the true causes, such hypotheses as being assumed, enable the motions to be calculated correctly from the principles of geometry, for the future as well as for the past ... These hypotheses need not be true nor even probable; if they provide a calculus consistent with the observations that alone is sufficient.[4]

Predictively successful mathematical models can be formulated for any periodic process. These models may have no explanatory value, however. Copernicus himself piled epicycle upon epicycle in Books II to VI of *De revolutionibus*. But neither Copernicus nor his geostatic theory rivals held that a planet is in Capricorn today in part because it is moving along an epicycle whose radius is a certain fraction of the radius of a deferent circle.

There also have been evaluative episodes within the history of science in which explanatory power is emphasized almost to the exclusion of predictive accuracy. Consider, for example, Darwin's "evolutionary histories" to account for the facts of biogeographical distribution. These histories "explain" by applying the principle of natural selection on the assumption that a number of conditions—an initial dispersion of the parent species, reproductive isolation, diversity of habitats, et al.—were realized in the past.

Evolutionary histories have explanatory value. But the arguments that express these histories invoke a multiply conditional premise of the form "(x)[(Ax & Bx & Cx & ...) $\supset$ Ψx]." Only multiply conditional predictions can be derived from such a premise. If "instance" a arises such that "Aa & Ba & Ca & ...," then Ψa follows. However, Darwin did not apply premises of this form in an effort to predict "new" biogeographical distributions. Instead he stressed the explanatory power of the theory of natural selection.

In some contexts, predictive accuracy and explanatory power are antithetical aims. When these aims do conflict, evaluative decisions are contingent upon judgments about their relative importance. Consider the case of the pressure-volume-temperature behavior of a gas near its critical point. If predictive accuracy is selected to be the more important aim, then the virial expansion is superior to Van der Waals' theory, and Van der Waals' theory is superior to ideal gas theory.

Ideal gas theory—$P V = k T$
Van der Waals' equation—$[P + a / V^2] (V=b) = k T$
Virial equation—$P V = A + B P + C P^2 + D P^3 + ...,$

where $A, B, C, ...$ are empirically determined, temperature-dependent coefficients that are specific to the gas in question. However, if explanatory power is selected to be the more important aim, then the virial equation is the least adequate of the three interpretations.

It would not be fatal to a foundationalist philosophy of science that there exist evaluative situations which cannot be resolved upon appeal to the basic foundational principles. But when foundational principles justify antithetical decisions in one and the same evaluative situation, there is a problem. Within McMullin's foundationalism, predictive accuracy and explanatory power appear to justify different comparative evaluative judgments in the context of interpreting the behavior of gases.

A supporter of McMullin's position on the fundamental aims of science might reply that a proper description of "the evaluative situation" requires specification of the appropriate aim. If the aim selected is predictive adequacy for P-V-T data, then select the appropriate virial expansion. On the other hand, if the aim is to understand gas behavior, then select Van der Waals' theory. These are not antithetical evaluative decisions, because there are two separate "evaluative situations."

However, the problem is reinstated if the aim selected is to achieve both predictive accuracy and explanatory power. If T_1 and T_2 are competing theories of a domain, and T_1 is superior on grounds of predictive accuracy, but inferior on grounds of explanatory power, then the conjunction of foundational principles provides no basis for evaluation. If instances in which predictive accuracy and explanatory power conflict are common, then the foundationalist needs to specify a principle that assigns relative importance to these fundamental aims.

The tension between predictive accuracy and explanatory power is reflected at a lower rung of the justificatory ladder in a tension between "agreement with observations" and "simplicity." Consider a set of observation reports on the relationship of properties A and B. A theory that implies that the data points be connected by straight lines maximizes agreement with observations. Given a set of data points, a competent mathematician can superimpose a power expansion upon them. The accuracy of fit (agreement with observations) is limited only by restrictions on the number of terms the mathematician is permitted to use in the expansion. However, a theory that states that "A ɔ 1/B" arguably would be more simple, even though no data point falls on this curve.

Since scientists usually prefer the smooth curve to the linear zigzag function, it is clear that they do not accept agreement with observations to be a sufficient condition of the acceptability of theories. Normally, they invoke considerations of simplicity as well.

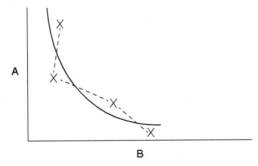

Figure 40 Relationship of Properties A and B.

The fact that hypothesis-appraisal usually reflects a tension between agreement-with-observations and simplicity provides support for the conclusion that conflicts between predictive accuracy and explanatory power are common as well. Since such conflicts are widespread, McMullin's "fundamental cognitive aims" are inadequate foundations for evaluative practice in science.

Methodological Rules as Foundational

Worrall on General Methodological Principles

John Worrall agreed with Shapere and Laudan that many evaluative standards have changed within the history of science. He insisted, nevertheless, that this is insufficient to establish that every standard is subject to change.

Worrall contrasted his methodological foundationalism with Laudan's naturalism. He maintained that there do exist general evaluative principles that are, and ought to be, in force at all stages of the development of science. He recommended the following candidates:

1. Theories should be tested against plausible rivals (if there are any).[1]
2. Non-*ad hoc* accounts should always be preferred to *ad hoc* ones (where both are available).[2]
3. Greater empirical support can legitimately be claimed for the hypothesis that a particular factor caused some effect if the experiment testing the hypothesis has been shielded against other possible causal factors.[3]

Worrall maintained that these principles underlie the practice of science in the same way that *modus ponens* underlies deductive logic. A person who accepts *p* and (p ⊃ q) but refuses to accept *q* (e.g., Lewis Carroll's Tortoise)[4] has opted out of the game of deductive logic. According to Worrall, a person who repudiates the inviolable principles at the top of the justificatory hierarchy likewise has opted out of the game of science.

Worrall insisted that the only way to escape historical relativism is to affirm trans-historical inviolable principles such as those above. Without such principles, appraisals of scientific evaluative practice lack genuine normative-prescriptive impact. The philosopher of science may counsel "if you wish your appraisals to conform to the standards of time *t*, then you ought to act in such-and-such ways." But such counsel is open to the objection that there is nothing special about time *t*, even if *t* is the present moment.

Worrall emphasized that without recourse to inviolable evaluative principles not themselves subject to justification by appeal to further principles, it will not be possible to demonstrate the rationality and progressive nature of science.

Laudan remained unconvinced. He noted that Worrall's principle that "theories should be tested against plausible rivals (if there are any)" is *not* a procedural principle applicable to all possible knowledge-contexts. On the contrary, it is a substantive principle whose applicability depends upon specific facts about the world. Laudan observed that it would be gratuitous to apply this principle to the generalization "All swans are white" in a universe known to contain a finite number of swans each of which has been examined and found to be white.

This criticism is not effective. Worrall called attention to the restriction he had placed upon the principle under consideration. The methodologist is required to test a theory against plausible rivals *only* in cases where there are genuine rivals. But Laudan's fictitious finite universe contains no such rival theory. To accept "All swans are black" to be a "rival" hypothesis under these conditions would be to deprive the concept "test" of all meaning.

On the other hand, Laudan is surely right to insist that "what sort of evidence constitutes a test of a theory and what does not" is learned within the history of science.[5] Whatever content is possessed by Worrall's principle is supplied by our changing understanding of "theory" and "test of a theory." Hence Laudan is correct that the principle is not a purely formal principle.

Laudan called attention to the principle of double-blind testing. Adoption of this principle in the twentieth century was a methodological innovation, an innovation based on discovery of the placebo effect. Laudan cited this development in support of his position that methodological rules arise out of our experience as we learn more about how to conduct inquiry.

Worrall conceded that some methodological principles have been developed (and some have been abandoned) in response to specific scientific achievements. He insisted, however, that there are some principles that are inviolable. Indeed, the double-blind principle is based, ultimately, on the basic inviolable principle that "theories should be tested against plausible rivals (if there are any)." The placebo hypothesis is always a viable competitor in the case of clinical trials.

Baysean Conditionalization

A number of philosophers of science have attributed foundational status to a subjectivist interpretation of a formula first developed by Thomas Bayes.[1] The subjectivist Bayesian position is that the degree of belief it is rational to attribute to hypothesis *h*, given additional evidence *e*, is given by the formula

$P(h/e)=P(e/h) P(h) / P(e)$

where "*P(h/e)*" is the posterior probability of hypothesis *h*, given evidence *e*, "*P(e/h)*" is the probability of the evidence, given the hypothesis and background assumptions, relative to the probability of the evidence on the background assumptions alone, "*P(h)*" is the prior probability of the hypothesis independent of *e*, and "*P(e)*" is the probability of the evidence.

The above formula stipulates the relationship between the degrees of rational belief in a hypothesis and additional evidence. The formula does not specify how to assign values to "*P(e)*," "*P(e/h)*," and "*P(h)*." But given these values, the value of "*P(h/e)*" is determined.

The relevant evidence in the application of the formula is "new evidence," evidence developed after assignment of a "prior probability" to the hypothesis. The formula is not directly applicable to "old evidence" of which the theorist was aware at the time that the prior probability was assigned. For "old evidence" e_0, "P $(e_0/h) = P (e_0) = 1$," and "P $(h/e_0) = P (h)$."

Bayesian theorists have tried various strategies to deal with the problem of "old evidence." An approach championed by Colin Howson and Peter Urbach is to introduce counterfactual degrees of belief.[2] They take the degree of support provided by old evidence to be that degree of support that would have been provided had the evidence in question been newly introduced. Of course, there is no way to replay the situation so that the old evidence becomes new evidence. Estimates of contrary-to-fact support are subject to debate and it is not clear how such debates can be resolved.

Daniel Garber has developed a different approach to the problem of old evidence. He suggested that old evidence may provide support for a hypothesis in virtue of a newly recognized entailment relationship.[3] If it is newly recognized that hypothesis *h* implies old evidence e_0, (i.e., h $\rightarrow$ e_0), then that evidence raises the degree of rational belief in *h*, viz., "P[h/e_0 & (h $\rightarrow$ e_0)] > P (h/e_0)."

It is important to realize that hypotheses do not imply evidence statements directly. The notation "*h* $\rightarrow$ e_0" is misleading. Additional premises are required, premises stating relevant conditions, and in most cases, auxiliary hypotheses as well. For instance, on Garber's approach, Newton's hypothesis of universal gravitational attraction receives support from old evidence expressed in Kepler's laws in virtue of the deductive relationship between the hypothesis and the laws. However, this deductive relationship involves premises other than the hypothesis of universal gravitational attraction. The additional premises state assumptions about relative masses ($m_{sun} >>> m_p$) and, in the case of the derivation of Kepler's Third Law, assumptions about non-interacting planets. Similarly, Copernicus' heliostatic hypothesis presumably receives support from the "old evidence" that the extent of retrograde motion of Mars is greater than that of Saturn. But the implication relation that provides this support holds only on specific assumptions about the orbital distances of the planets and their relative speeds in these orbits.

There are difficulties then in extending the Bayesian approach to old evidence. But even if attempts to account for the evidential role of old evidence are judged unsatisfactory, the Bayesian formula, restricted in application to new evidence, is still a candidate for foundational status.

The formula specifies the way in which what it is rational to believe changes with the accumulation of additional evidence. Different theorists most likely will assign different probabilities to a hypothesis initially. But repeated applications of the formula to an increasing body of evidence will lead to convergence upon a particular posterior probability for the hypothesis. Bayesians typically illustrate this learning process by such examples as the selection of balls from an urn. Initially theorists may have widely different expectations about the number of red balls in an urn. But as balls are withdrawn, examined, replaced, randomized, and withdrawn again, the observers' calculated posterior probabilities will converge upon the true value.

Of course, scientific research seldom resembles the repetitive withdrawal of objects from a container. It is clear that withdrawal of a non-red ball reduces the probability calculated for the percentage of red balls in the urn. However, it is not clear that a case of *prima facie* negative evidence invariably counts against a scientific hypothesis.

In any given instance, the negative evidence may be accommodated, using the formula, by adjusting the value of the prior probability of the hypothesis and leaving unchanged the posterior probability. Richard Miller has argued that a recalculation of *a priori* probability is sometimes the better response to negative evidence.[4] This is what Darwin did when confronted with evidence about the absence of fossil remains of transitional forms in the tree of descent. Darwin urged that we revise our expectations about the likelihood of fossilization. He did not lower his estimate of probability that the operation of natural selection created a tree of descent. Nor did Galileo decrease the probability of his hypothesis that telescopic evidence of the heavens is reliable, given evidence that the telescope reduces the apparent size of stars while enlarging the apparent size of the planets (and terrestrial objects). Galileo suggested instead that our prior expectations be revised to account for the effect of the "adventitious rays" from the stars that affect the naked eye but are removed by the telescope.

The Bayesian formula is a stipulative definition of "the change in degree of belief it is rational to accord a hypothesis given additional evidence." The methodologist would like to use the formula to unambiguously assign degrees of rational belief to a hypothesis as evidence accumulates. However, the formula itself it not a foundational evaluative standard. It does not discriminate between the recalculation of the probability of a hypothesis on additional evidence and the revision of the assumed prior probability so as to cancel the impact of the evidence. The methodologist will require further instruction on how to assign values within the formula in specific evaluative contexts. In particular, she needs to know under what conditions it is permissible to revise prior probabilities.

Descriptivism Versus Foundationalism

The normative naturalist positions discussed above either (1) import trans-historical evaluative principles to warrant normative claims (Neurath, Quine, Kitcher, Laudan), or (2) fail to establish normative-prescriptive status for the evaluative standards that are developed within the history of science (Burian, Shapere). Given the failure to create a third alternative, descriptivism and foundationalism remain the principal contending philosophies of science.

The Case for Descriptivism

Descriptivism and foundationalism are both viable positions on the philosophy of science. No philosopher should object to the descriptivist project to uncover the standards presupposed in actual evaluative practice. However, descriptivism is subject to challenge if its practitioners maintain that there is nothing further to be done in the philosophy of science. Foundationalists insist that philosophy of science ought be a search for principles whose applications have normative-prescriptive force.

It might seem that to argue on behalf of one of these alternatives is to beg the question against the other. To argue for normative-prescriptive intent for the discipline is to exclude pure descriptivism, and to argue for exclusively descriptive intent is to exclude normative-prescriptive considerations.

The situation is not quite so desperate, however. Friedrich Waismann's description of how philosophical disputes are resolved is appropriate in this context. According to Waismann, the philosopher, like an attorney at the bar, seeks to build a case. She does so by emphasizing the disadvantages of the opponent's position, and by submitting an alternative position not subject to these disadvantages. The aim of the discussion, in both a judicial proceeding and a philosophical debate, is to gain a favorable verdict.

Waismann emphasized that

in philosophy there are no proofs; there are no theorems—and there are no questions that can be decided Yes or No.[1]

This is not to say that there are no good arguments in philosophy. There are. It's just that these arguments are not forcing.

One can build a case for the descriptivist option. One might begin by conducting a survey of the attitudes of practicing scientists. The consensus presumably would be that evaluative decisions ought be left to scientists themselves. This is a reasonable position in the face of the historical record. This record does not feature scientific achievements attributable to the normative advice of philosophers. Descriptivism is the appropriate response in this situation. Philosophers who are far removed from the cutting edge of research are ill-equipped to pass judgment on theory-choice and explanatory adequacy.

Normative-prescriptive theorists nevertheless presume to legislate proper scientific methodological and evaluative practice. Descriptivism is the more modest position. The descriptivist is an exhibitor and not an arbitrator. If modesty is a virtue, then descriptivism is the more virtuous position.

Moreover, if one looks at the history of the philosophy of science, the normal sequence has been, first, new developments in science, and second, philosophical analyses of the methodological and evaluative implications of these developments. A good example is the sequence that proceeds from Einstein's formulation of special relativity theory to Bridgman's operationalist commentary on this achievement. Given that philosophy of science normally is a *reaction* to developments in science, the descriptivist may argue that it is implausible to assign a *leading* normative role to principles developed within a philosophy of science.

Of course, this argument from temporal priority is not forcing. It might be the case that true and inviolable principles have been uncovered from analyses of achievements in science, and that these principles ought be applied subsequently in all similar evaluative contexts. Although scientific theories replace one another over time (all theories are born false), some methodological principles, once discovered, might be applicable at all times.

Clearly this is not the case for every methodological principle. Laudan and others have called attention to changes at the methodological level. Some principles, once accepted, have been abandoned. An example is the pre-quantum-theory epistemological principle that a full explanation of an experimental result include both a spatio-temporal description of an evolving system and a causal analysis of that evolution.

Perhaps the strongest criticism of foundationalism is that the principles proposed to be foundational have only an inconsequential normative-predictive impact. In a given evaluative context, there often are numerous options that are consistent with the foundational standards. To require that acceptable scientific theories be internally consistent, have observable consequences, and triumph in tests against available rival theories is to place only minimal restrictions on theory-choice. Given this limited normative-prescriptive import, one might as well opt for the relative security of the descriptivist standpoint.

The Case for Foundationalism

The foundationalist may agree that requirements such as internal consistency and operational links to observables fail to decide the issue among theories that fulfill these requirements, but insist, nevertheless, that these principles state necessary conditions of acceptability for scientific theories. No theory that is inconsistent, or that lacks links to observables, is acceptable. Moreover, the foundationalist may maintain that the methodological requirement that theories be tested against available rivals is a sound principle even if disputes arise about what constitutes a test in particular cases.

One initially plausible argument is based on an appeal to the theory of organic evolution. One may argue that a principle is justified because its applications have proved to be adaptive for *homo sapiens*. Arguments of this type are not decisive, however. Three types of objections may be raised. The first objection, to parody Feyerabend, is "what is so great about *homo sapiens*?" Granted that a type of response to environmental pressures had proved adaptive for *homo sapiens*, why should we take the continued success of *homo sapiens* to be the measure of evolutionary development?

The second objection is that the evolutionary defense involves an element of circularity. Commitment to intra-theoretical consistency, for example, is justified by appeal to evolutionary theory, which theory is itself justified, in part, by appeal to intra-theoretical consistency.

The third objection is perhaps the most damaging. "Evolutionary success" requires both adaptation and retention of adaptability. It always is possible that adaptation in a specific context has been achieved in such a way as to diminish the capacity to respond creatively to further changes in environmental conditions. Adaptation may have been purchased at the expense of a loss of adaptability.

It is possible, at least, that this is the situation with respect to methodological principles hitherto accepted as foundational. The evolutionist cannot prove that adaptability has been retained. That adaptability has been retained can be established only by reference to what happens at subsequent times. Since the above objections have some force, the evolutionary argument to justify the inviolable status to methodological principles is not effective.

Perhaps a burden-of-proof shift is useful at this point. The opponent is invited to cite historical episodes that violate one or more of the above requirements but nevertheless were progressive. Of course, there are competing theories about what counts as progress in science. The effectiveness of the burden-of-proof strategy depends on antecedent agreement about scientific progress. But even if agreement is not forthcoming, the foundationalist may restate the challenge by reference to the historical record itself. The opponent is invited to cite developments (since, say, 1700) that violate one of the foundational principles but were accepted by a majority of scientists practicing in the relevant field for a specified period of time.

Failure of the opponent to cite historical episodes that violate supposedly foundational principles would strengthen the foundationalist position. But such failure would not establish that these principles *ought be held* to be inviolable. That certain principles have been ubiquitous within scientific evaluative practice does not entail that these principles should be endorsed as governing future evaluative decisions. However, the foundationalist may point out that methodologists do not debate the status of intra-theoretical consistency or the operational requirement in the way in which they debate the status of undesigned scope or complementarity. The last resort for the foundationalist is to affirm, with Worrall, that

> ultimately we must stop arguing and "dogmatically" assert certain basic principles of rationality.[1]

Notes

Chapter 1: Logical Reconstructionism

Sources of Logical Reconstructionism (pp. 5–19)

1. Norman R. Campbell, *Foundations of Science*, formerly *Physics: The Elements*, 1919 (New York: Dover, 1957), 1–12.
2. A survey of views about scientific method is presented in John Losee, *A Historical Introduction to the Philosophy of Science,* 4th edn. (Oxford: Oxford University Press, 2001).
3. Paul K. Feyerabend, "Philosophy of Science: A Subject with a Great Past," in *Historical and Philosophical Perspectives on Science, Minnesota Studies in the Philosophy of Science, vol. V*, ed. R. Stuewer (Minneapolis: University of Minnesota Press, 1970), 181.
4. Representative presentations of the logical positivist position are: A. J. Ayer, *Logic, Truth and Knowledge* (New York: Dover, 1944); Rudolf Carnap, "The Elimination of Metaphysics through Logical Analysis of Language," in *Logical Positivism*, ed. A. J. Ayer (Glencoe, IL: Free Press, 1959); Moritz Schlick, "The Turning Point in Philosophy," in *Logical Positivism. Logical Positivism* contains a useful bibliography of positivist writings.
5. Ludwig Wittgenstein, *Tractatus Logico-Philosophicus*, trans. D. F. Pears and B. F. McGuinness (New York: Humanities Press, 1961).
6. Ibid., 41.
7. Rudolf Carnap, "Testability and Meaning," *Phil. Sci. 3* (1936) and *4* (1937). Reprinted in *Readings in the Philosophy of Science*, ed. H. Feigl and M. Brodbeck (New York: Appleton-Century-Crofts, 1953).
8. Carnap, "The Elimination of Metaphysics," 63.
9. Carnap, "Testability and Meaning," in *Readings in the Philosophy of Science*, 52–6.
10. Carnap, "Logical Foundations of the Unity of Science," in *International Encyclopedia of Unified Science, vol. 1, no. 1*. Reprinted in *Readings in Philosophical Analysis*, ed. H. Feigl and W. Sellars (New York: Appleton-Century-Crofts, 1949), 416–18.
11. Carnap, "The Methodological Character of Theoretical Concepts," in *Minnesota Studies in the Philosophy of Science, vol. I*, ed. H. Feigl and M. Scriven (Minneapolis: University of Minnesota Press, 1956), 39.
12. Werner Heisenberg, *Physics and Philosophy* (New York: Harper & Brothers, 1958), 46.

13. Heisenberg, *Physics and Philosophy*, 44; Niels Bohr, "Quantum Physics and Philosophy," in *Essays 1958–1962 on Atomic Physics and Human Knowledge* (New York: Interscience, 1963), 1–7.

14. Max Born, *Atomic Physics* (New York: Hafner, 1946), 92–100. See also Born, *Physics in My Generation* (New York: Pergamon Press, 1956).

15. Ernst Mach, *The Science of Mechanics* (1883), trans. T. J. McCormack (Chicago: Open Court, 1960), 265–71.

16. Ibid., 271–97.

17. Henri Poincare, *Science and Hypothesis*, trans. G. B. Halsted (New York: The Science Press, 1905), 73–8.

18. Pierre Duhem, *The Aim and Structure of Physical Theory*, 2nd edn. (1914), trans. P. P. Wiener (New York: Atheneum, 1968), 128–9.

19. Albert Einstein, *Relativity* (New York: Crown, 1961), 22–6.

20. P. W. Bridgman, *The Nature of Physical Theory* (Princeton, NJ: Princeton University Press, 1936), 10.

21. Gustav Bergmann, "Sense and Nonsense in Operationalism," in *The Validation of Scientific Theories*, ed. P. Frank (Boston: Beacon Press, 1954), 43.

22. Henry Margenau, *The Miracle of Existence* (Woodbridge, CT: Ox Bow Press, 1984), 53–4.

23. Bridgman, *The Nature of Physical Theory*, 66; *The Logic of Modern Physics* (New York: Macmillan, 1960), 54–6.

24. Bridgman, *The Logic of Modern Physics*, 28–9.

25. Ibid., 9.

26. Bergmann, "Sense and Nonsense in Operationalism," 50.

27. Carl Hempel, "A Logical Appraisal of Operationalism," in *The Validation of Scientific Theories*, 56.

28. R. B. Lindsay, "A Critique of Operationalism in Physics," *Phil. Sci. 4* (1937), 456.

29. Bridgman, *The Way Things Are* (Cambridge, MA: Harvard University Press, 1959), 51.

30. Bridgman, *Reflections of a Physicist* (New York: Philosophical Library, 1950), 9.

31. Hans Reichenbach, T*he Rise of Scientific Philosophy* (Berkeley: University of California Press, 1951), 231.

Patterns of Explanation (pp. 19–29)

1. Carl. G. Hempel and Paul Oppenheim, "Studies in the Logic of Explanation," *Phil. Sci. 15* (1948), 135–75; reprinted in Hempel, *Aspects of Scientific Explanation* (New York: Free Press, 1965), 245–95.

2. Ibid., 273. See also "Deductive-Nomological vs. Statistical Explanation," in *Minnesota Studies in the Philosophy of Science, vol. III,* ed. H. Feigl and G. Maxwell

(Minneapolis: University of Minnesota Press, 1956), 102–3; Hempel, *Aspects of Scientific Explanation*, 338.

3. Rolf Eberle, David Kaplan, and Richard Montague, "Hempel and Oppenheim on Explanation," *Phil. Sci. 28* (1961), 419–28.

4. Kaplan, "Explanation Revisited," *Phil. Sci. 28* (1961), 430–1.

5. Hempel and Oppenheim, "Studies in the Logic of Explanation," 250–1.

6. Hempel, *Aspects of Scientific Explanation*, 382.

7. Hempel and Oppenheim, "Studies in the Logic of Explanation," 249.

8. Arturo Rosenblueth, Norbert Wiener, and Julian Bigelow, "Behavior, Purpose and Teleology," *Phil. Sci. 10* (1943), 18–24.

9. Norbert Wiener, *The Human Use of Human Beings: Cybernetics and Society* (Garden City, NY: Doubleday Anchor, 1956), 33.

10. Rosenblueth, Wiener, and Bigelow "Behavior, Purpose and Teleology," 18.

11. Richard Taylor, "Comments on a Mechanistic Conception of Purposefulness," *Phil. Sci. 17* (1950), 310–17.

12. Rosenblueth and Wiener, "Purposeful and Non-Purposeful Behavior," *Phil. Sci. 17* (1950), 318–26.

13. Ibid., 324.

14. Ibid., 325.

15. Ibid., 320.

16. Taylor, "Purposeful and Non-Purposeful Behavior: A Rejoinder," *Phil. Sci. 17* (1950), 327–32.

17. Ibid., 330.

18. Rosenblueth, Wiener, and Bigelow, "Behavior, Purpose and Teleology," 18; Rosenblueth and Wiener, "Purposeful and Non-Purposeful Behavior," 324.

19. Ernest Nagel, *The Structure of Science* (New York: Harcourt, Brace & World, 1961), chapter 12, section 1.

20. Ibid., 409–22.

21. Ibid., 405–6.

Laws and Confirmation (pp. 29–45)

1. Norman Robert Campbell, *Foundations of Science* (New York: Dover, 1957), 89.

2. Ibid., 39–40.

3. Richard B. Braithwaite, *Scientific Explanation* (Cambridge: Cambridge University Press, 1953), 302.

4. Ernest Nagel, *The Structure of Science* (New York: Harcourt, Brace & World, 1961), 56–67.

5. Ibid., 71–2.

6. Arthur Pap, *An Introduction to the Philosophy of Science* (Glencoe, IL: Free Press, 1962), 304–5.

7. Carl Hempel, "Studies in the Logic of Confirmation," *Mind 54* (1945), 1–26; 97–121; reprinted in Hempel, *Aspects of Scientific Explanation* (New York: Free Press, 1965), 3–46.

8. Hempel, *Aspects of Scientific Explanation*, 41.

9. Jean Nicod, "The Logical Problem of Induction," in *Geometry and Induction*, trans. J. Bell and M. Woods (London: Routledge & Kegan Paul, 1969), 189.

10. Hempel, "Studies in the Logic of Confirmation," in *Aspects of Scientific Explanation*, 13.

11. Ibid., 18–19.

12. Ibid., 14–15.

13. Ibid., 20.

14. Hempel, *Aspects of Scientific Explanation*, 31–3.

15. Ibid., 34.

16. Ibid., 37.

17. Ibid.

18. Ibid.

19. Rudolf Carnap, *Logical Foundations of Probability* (Chicago: University of Chicago Press, 1950), 473–4.

20. Ibid., 477.

21. Hempel, "Postscript (1964) on Confirmation," in *Aspects of Scientific Explanation*, 48–50.

22. Carnap, *Logical Foundations of Probability*.

23. Carnap, *The Continuum of Inductive Methods* (Chicago: University of Chicago Press, 1952); *The Logical Structure of the World* (London: Routledge & Kegan Paul, 1967).

24. Carnap, *Logical Foundations of Probability*, 289.

25. Ibid.

26. Ibid., 564–5.

27. Ibid., 571.

28. Ibid., 572.

29. Ibid., 572–3.

30. Ibid., 576.

Theories as Interpreted Axiom Systems (pp. 46–76)

1. Norman R. Campbell, *Foundations of Science*, formerly *Physics: The Elements*, 1919 (New York: Dover, 1957).

2. Ibid., 122.

3. Pierre Duhem, *The Aim and Structure of Physical Theory* (1906), trans. P. P. Wiener (New York: Atheneum, 1962), 19–21.

4. Campbell, *Foundations of Science*, 104.

5. Ibid., 129.

6. Ibid.

7. Gerd Buchdahl, "Theory Construction: The Work of Norman Robert Campbell," *ISIS 55* (1964), 158.

8. Campbell, *Foundations of Science*, 146.

9. Ibid., 143.

10. Ibid., 141.

11. Rudolf Carnap, "Foundations of Logic and Mathematics" (1939), in *International Encyclopedia of Unified Science, vol. I, part 1*, ed. O. Neurath, R. Carnap, and C. Morris (Chicago: University of Chicago Press, 1955), 202.

12. Philipp Frank, "Foundations of Physics," in *International Encyclopedia of Unified Science, vol. I, part 2*; Carl Hempel, "Fundamentals of Concept Formation in Empirical Science," in *International Encyclopedia of Unified Science, vol. I, no. 7.*

13. Carnap, "Foundations of Logic and Mathematics," 204.

14. Ibid., 205.

15. Ibid., 207.

16. Hempel, "Fundamentals of Concept Formation in Empirical Science," 21.

17. Ibid., 39.

18. Richard B. Braithwaite, *Scientific Explanation* (Cambridge: Cambridge University Press, 1955).

19. Ibid., 51–2, 88–93.

20. Ibid., 80.

21. Ibid., 88–92.

22. Henry Margenau, *The Miracle of Existence* (Woodbridge, CT: Ox Bow Press, 1984), 52.

23. Ibid.

24. Ibid.

25. Ibid., 53–4.

26. Ibid., 54.

27. Ludwig Wittgenstein, *Philosophical Investigations* (New York: Macmillan, 1952), part I, ∮ 202, ∮ 206–8.

28. Margenau, "Metaphysical Elements in Physics," in *Physics and Philosophy: Selected Essays* (Dordrecht: Reidel, 1978), 116.

29. Ibid., 101–2.

30. Ibid., 102.

31. Ibid., 103.

32. Ibid., 104.

33. Margenau, *The Nature of Physical Reality* (New York: McGraw-Hill, 1950), 81–100.
34. Ibid., 299.
35. Philipp Frank, *Philosophy of Science* (Englewood Cliffs, NJ: Prentice-Hall, 1957), 356.
36. Ibid., 130–3.
37. Heinrich Hertz, *Electric Waves*, trans. D. E. Jones (London: Macmillan, 1900), 20–8.
38. Frank P. Ramsey, *The Foundations of Mathematics* (Paterson, NJ: Littlefield, Adams & Co., 1960), 237–55; Moritz Schlick, *Gesammelte aufsetze* (Vienna: Gerold, 1938), 67–8.
39. Gilbert Ryle, *The Concept of Mind* (New York: Barnes & Noble, 1949), 121.
40. Stephen Toulmin, *The Philosophy of Science* (New York: Harper & Brothers, 1953), 70.
41. William Craig, "On Axiomatizability within a System," *J. Symbolic Logic 18* (1953), 30–2.
42. Craig, "Replacement of Auxiliary Expressions," *Phi. Rev. LXV* (1956), 38–53; Hempel. "The Theoretician's Dilemma," in *Minnesota Studies in the Philosophy of Science, vol. II*, ed. H. Feigl, M. Scriven, and G. Maxwell (Minneapolis: University of Minnesota Press, 1958), 37–98; reprinted in Hempel, *Aspects of Scientific Explanation* (New York: Free Press, 1965), 173–226; Israel Scheffler, "Prospect of a Modest Empiricism, II," *Rev. Metaphysics X* (1957), 619; *Anatomy of Inquiry* (New York: Bobbs-Merrill, 1963), 193–203.
43. Ernest Nagel, *The Structure of Science* (New York: Harcourt, Brace & World, 1961) 129–52.
44. Ibid., 138.
45. Ibid., 152.

Reduction and the Growth of Science (pp. 76–78)

1. Ernest Nagel, *The Structure of Science* (New York: Harcourt, Brace & World, 1961), 336–7.
2. Ibid., 335–9.
3. Ibid., 345–66.
4. Ibid., 338–45.
5. Niels Bohr, *Atomic Theory and the Description of Nature* (Cambridge: Cambridge University Press, 1961), 35–9.
6. Ernest Hutten, *The Language of Modern Physics* (New York: Macmillan, 1956), 166–7.

Karl Popper on Justification and Discovery (pp. 79–85)

1. Karl Popper, *The Logic of Scientific Discovery*, 1934 (New York: Basic Books, 1959).
2. Ibid., 49–50.
3. Ibid., 54.
4. Ibid., 111.

5. Ibid., 265–9.

6. Ibid., 276.

7. Carl Hempel, *Philosophy of Natural Science* (Englewood Cliffs, NJ: Prentice-Hall, 1966), 38.

8. John F. W. Herschel, *A Preliminary Discourse on the Study of Natural Philosophy*, 1830 (New York: Johnson Reprint, 1960), 170–2.

9. Popper, *The Logic of Scientific Discovery*, 269.

Chapter 2: Orthodoxy Under Attack

Doubts about the Two-Tier View of Scientific Language (pp. 87–108)

1. Herbert Feigl, "Existential Hypotheses," *Phil. Sci. 17* (1950).

2. Ibid., 41.

3. Ibid., 192.

4. Carl Hempel, "A Note on Semantic Realism," *Phil. Sci. 17* (1950). 169–73; Ernest Nagel, "Science and Semantic Realism," *Phil. Sci. 17* (1950), 174–81.

5. Hempel, "A Note on Semantic Realism," 172.

6. C. W. Churchman, "Logical Reconstructionism," *Phil. Sci. 17* (1950), 164.

7. Feigl, "Logical Reconstruction, Realism and Pure Semiotic," *Phil. Sci. 17* (1950), 186–95.

8. Ibid., 195.

9. Mary Hesse, "Theories, Dictionaries, and Observation," *Brit. J. Phil. Sci. IX* (1958), 12–28.

10. Peter Alexander, "Theory-Construction and Theory-Testing," *Brit. J. Phil. Sci. IX* (1958), 29–38.

11. Ibid., 35.

12. Ibid.

13. Peter Achinstein, *Concepts of Science* (Baltimore, MD: The Johns Hopkins Press, 1968), 168.

14. Ibid., 157–201.

15. N. R. Hanson, *Patterns of Discovery* (Cambridge: Cambridge University Press, 1958), 58–62.

16. Ibid., 59.

17. Gilbert Ryle, *Dilemmas* (Cambridge: Cambridge University Press, 1954), 90–1.

18. Achinstein, *Concepts of Science,* 183.

19. Ibid., 199.

20. Paul Feyerabend, "An Attempt at a Realistic Interpretation of Experience," *Proc. Arist. Soc. 58* (1958), 160–2.

21. Ibid., 163.

22. Willard van Orman Quine, "Two Dogmas of Empiricism," in *From a Logical Point of View* (Cambridge: Harvard University Press, 1953), 42–3.

23. Pierre Duhem, *The Aim and Structure of Physical Theory*, 1914 (New York: Atheneum, 1962), 182–218.

24. Quine, "Two Dogmas of Empiricism," 44.

25. Hans Reichenbach, *Philosophical Foundations of Quantum Mechanics* (Berkeley: University of California Press, 1948), 144–70.

26. Quine, "Two Dogmas of Empiricism," 43.

27. Feigl, "Some Major Issues and Developments in the Philosophy of Science of Logical Empiricism," in *Minnesota Studies in the Philosophy of Science, vol. I*, ed. H. Feigl and M. Scriven (Minneapolis: University of Minnesota Press, 1956), 6–7.

28. H. P. Grice and P. F. Strawson, "In Defense of a Dogma," *Phil. Rev.* 65 (1956), 143.

29. Grover Maxwell, "Meaning Postulates in Scientific Theories," in *Current Issues in the Philosophy of Science*, ed. H. Feigl and G. Maxwell (New York: Holt, Rinehart & Winston, 1961), 169–83; "The Necessary and the Contingent," in *Minnesota Studies in the Philosophy of Science, vol. III*, ed. by H. Feigl and G. Maxwell (Minneapolis: University of Minnesota Press, 1962), 398–404.

30. Hilary Putnam, "The Analytic and the Synthetic," in *Minnesota Studies in the Philosophy of Science, vol. III*, 359–60.

31. Ibid., 375.

32. Ibid., 385–6.

33. Adolf Grunbaum, "The Duhemian Argument," *Phil. Sci.* 27 (1960). 75–87; "Geometry, Chronometry, and Empiricism," in *Minnesota Studies in the Philosophy of Science, vol. III*, 405–526; "The Falsifiability of Theories: Total or Partial?" in *Boston Studies in the Philosophy of Science, vol. I*, ed. M. Wartofsky (Dordrecht: Reidel, 1963), 178–95.

34. Grunbaum, "The Falsifiability of Theories," 181.

35. Albert Einstein, "Reply to Criticisms," in *Albert Einstein: Philosopher-Scientist*, ed. P. A. Schilpp (New York: Tudor, 1951), 676–8.

36. Grunbaum, "The Falsifiability of Theories," 189.

37. Ibid., 187–8.

38. Israel Scheffler, *Science and Subjectivity* (Indianapolis, IN: Bobbs-Merrill, 1967), 8–10.

39. Ibid., 60.

40. Mary Hesse, "Review of Israel Scheffler's *Science and Subjectivity*," *Brit. J. Phil. Sci.* 19 (1968–69), 177.

Doubts about Explanation (pp. 108–123)

1. Michael Scriven, "Truisms as the Grounds for Historical Explanations," in *Theories of History*, ed. P. Gardiner (Glencoe, IL: Free Press, 1959), 443–75; "Explanations and Prediction in Evolutionary Theory," *Science 130*, 477–82; "Explanations, Predictions

and Laws," in *Minnesota Studies in the Philosophy of Science, vol. III*, ed. H. Feigl and G. Maxwell (Minneapolis: University of Minnesota Press, 1962), 170–230.

2. Scriven, "Explanations, Predictions and Laws," 197.

3. Scriven, "Truisms as the Grounds for Historical Explanations," 456; "Explanations, Predictions and Laws," 198.

4. Scriven, "Truisms as the Grounds for Historical Explanations," 456.

5. Ibid., 457.

6. Hempel, *Aspects of Scientific Explanation*, 362.

7. See David Lack, *Darwin's Finches* (New York: Harper Torchbooks, 1961), 17–22.

8. Hempel, *Aspects of Scientific Explanation*, 373.

9. Wesley Salmon, "Why Ask 'Why'? An Inquiry concerning Scientific Explanation," *Proc. Arist. Soc. 6* (1978), 689; reprinted in *Scientific Knowledge*, ed. J. Kourany (Belmont, CA: Wadsworth, 1987), 56.

10. William Dray, *Laws and Explanations in History* (Oxford: Clarendon Press, 1957), 61.

11. Rom Harre, *The Principles of Scientific Thinking* (London: Macmillan, 1970), 20.

12. Harre, "Metaphor, Model and Mechanism," *Proc. Arist. Soc. 60*, (1959–60), 102–3.

13. Hempel, "Deductive-Nomological vs. Statistical Explanations," in *Minnesota Studies in the Philosophy of Science, vol. III*, 109–10.

14. Ibid., 109.

15. Carl G. Hempel and Paul Oppenheim, "Studies in the Logic of Explanation," in Hempel, *Aspects of Scientific Explanation* (New York: Free Press, 1965), 249.

16. See, for instance, Pierre Duhem, *To Save the Phenomena*, trans. E. Doland and C. Maschler (Chicago: University of Chicago Press, 1969); Stephen Toulmin, *Foresight and Understanding* (New York: Harper & Row, 1961).

17. Hempel, *Aspects of Scientific Explanation*, 374–5.

18. Nicholas Rescher, "On Prediction and Explanation," *Brit. J. Phil. Sci. 8* (1958), 282.

19. N. R. Hanson, "On the Symmetry between Explanation and Prediction," *Phil. Rev. 68* (1959), 354.

20. Israel Scheffler, "Explanation, Prediction and Abstraction," *Brit. J. Phil. Sci. 7* (1957), 293–309.

21. Scriven, "Explanation and Prediction in Evolutionary Theory," 477–82; "Explanations, Predictions, and Laws," 170–230; "The Temporal Asymmetry of Explanations and Predictions," *Philosophy of Science: The Delaware Seminar, vol. 1*, ed. B. Baumrin (New York: Interscience, 1963), 97–105.

22. Scriven, "Explanations, Predictions, and Laws," 182–4.

23. Scriven, "Explanation and Prediction in Evolutionary Theory," 480; a similar argument was presented by S. Barker, "The Role of Simplicity in Explanation," in *Current Issues in the Philosophy of Science*, ed. H. Feigl and G. Maxwell (New York: Holt, Rinehart & Winston, 1961), 271.

24. Scriven, "Explanation and Prediction in Evolutionary Theory," 477–82.

25. Adolf Grunbaum, "Temporally Asymmetric Principles, Parity between Explanation and Prediction, and Mechanism and Teleology," *Phil. Sci. 29* (1962), 146–70; reprinted in *Philosophy of Science: The Delaware Seminar, Vol 1*, 57–96.

26. Ibid., 166.

27. Ibid.

28. Scriven, "The Temporal Asymmetry of Explanations and Predictions," 103.

29. Hempel, *Aspects of Scientific Explanation*, 371.

30. Hempel, "Reasons and Covering Laws in Historical Explanation," in *Philosophy and History*, ed. S. Hook (New York: New York University Press, 1963), 148–9.

31. Paul Dietl, "Paresis and the Alleged Asymmetry between Explanation and Prediction," *Brit. J. Phil. Sci. 17* (1966), 313–18.

32. Hempel, *Aspects of Scientific Explanation*, 369–70.

33. Scriven, "Explanation and Prediction in Evolutionary Theory," 479–80.

New Problems about Confirmation and Justification (pp. 123–138)

1. Nelson Goodman, *Fact, Fiction and Forecast*, 2nd edn. (Indianapolis: Bobbs-Merrill, 1965).

2. Ibid., 74.

3. Ibid., 79–80.

4. Ibid., 78–9.

5. Ibid., 94.

6. Ibid., 99.

7. Carl Hempel, "Postscript (1964) on Confirmation" in *Aspects of Scientific Explanation* (New York: Free Press, 1965), 51.

8. Ibid., 51.

9. Karl Popper, *Conjectures and Refutations* (New York: Basic Books, 1962), 36–7, 241–2; "The Aim of Science," in *Objective Knowledge* (Oxford: Oxford University Press, 1972), 192–3.

10. Elie Zahar, "Why Did Einstein's Programme Supersede Lorentz's?" *Brit. J. Phil. Sci. 24* (1973), 103.

11. Gerald Holton, "Einstein, Michelson, and the 'Crucial' Experiment," *ISIS 60* (1969), 133–97.

12. Alan Musgrave, "Logical versus Historical Theories of Confirmation," *Brit. J. Phil. Sci. 25* (1974), 1–23.

13. Imre Lakatos, "Changes in the Problem of Inductive Logic," in *Inductive Logic*, ed. I. Lakatos (Amsterdam: North-Holland, 1968), 315–417.

14. Lakatos, "Falsification and the Methodology of Scientific Research Programmes" in *Criticism and the Growth of Knowledge*, ed. I. Lakatos and A. Musgrave (Cambridge, Cambridge University Press, 1970), 116–19.

15. Karl Popper, "Degree of Confirmation," *Brit. J. Phil. Sci.* 5 (1954), 143–9.
16. John G. Kemeny, "Review of K. R. Popper's 'Degree of Confirmation,'" *J. Symbolic Logic XX* (1955), 304.
17. Y. Bar-Hillel, "Comments on 'Degree of Confirmation' by Professor K. R. Popper," *Brit. J. Phil. Sci. VI* (1955), 155–7.
18. Rudolf Carnap, *Logical Foundations of Probability* (Chicago: University of Chicago Press, 1950), 164.
19. Popper, "'Content' and 'Degree of Confirmation': A Reply to Dr. Bar-Hillel," *Brit. J. Phil. Sci. VI* (1955), 157–63.
20. Ibid., 160.
21. Carnap, "Remarks on Popper's Note on Content and Degree of Confirmation," *Brit. J. Phil. Sci. VII* (1956), 244.
22. Lakatos, "Changes in the Problem of Inductive Logic," 315–417.
23. Ibid., 332–3.
24. Ibid., 364.
25. Ibid., 372.
26. Alex C. Michalos, *The Popper–Carnap Controversy* (The Hague: Martinus Nijhoff, 1971), 78.
27. Lakatos, "Changes in the Problem of Inductive Logic," 368.
28. Popper, "Degree of Confirmation," 147.
29. Arthur W. Burks, "Justification in Science," in *Academic Freedom, Logic and Religion*, ed. M. White (Philadelphia: University of Pennsylvania Press, 1953), 112.
30. Ibid., 120.
31. Frederick L. Will, "The Justification of Theories," *Phil. Rev.* 64 (1955), 374–7.
32. Ibid., 381.
33. Ibid., 382.

Doubts about the Orthodox View of Theories (pp. 138–144)

1. Frederick Suppe, "The Search for Philosophic Understanding of Scientific Theories," in *The Structure of Scientific Theories*, ed. F. Suppe (Urbana: University of Illinois Press, 1974), 221–30; "Theory Structure," in *Current Research in Philosophy of Science*, ed. P. Asquith and H. Kyburg (Ann Arbor, MI: Philosophy of Science Association, 1979), 317–18. This essay also discusses "non-statement" positions defended by Beth, Suppes, Stegmuller, and van Fraassen.
2. The sentence–proposition distinction is discussed in S. Gorovitz and R. G. Williams, *Philosophical Analysis* (New York: Random House, 1963), chapter IV.
3. Suppe, "The Search for Philosophic Understanding of Scientific Theories," 222.
4. Ibid., 223–6.
5. Ernest Nagel, *The Structure of Science* (New York: Harcourt, Brace & World, 1961), 33.

6. Pierre Duhem, *The Aim and Structure of Physical Theory*, 1914, trans. P. Wiener (New York: Atheneum, 1962), 32.

7. Wilfrid Sellars, "The Language of Theories," in *Current Issues in the Philosophy of Science*, ed. H. Feigl and G. Maxwell (New York: Holt, Rinehart & Winston, 1961), 71–2; reprinted in *Readings in the Philosophy of Science*, ed. B. Brody (Englewood Cliffs, NJ: Prentice-Hall, 1970), 348.

8. Ibid., 74.

9. Marshall Spector, "Models and Theories," *Brit. J. Phil. Sci. 16* (1965–66), 121–42; reprinted in *Readings in Philosophy of Science*, 276–93.

10. Ibid., 286.

11. Ibid., 284.

12. See, for instance, S. L. Soo, *Analytical Thermodynamics* (Englewood Cliffs, NJ: Prentice-Hall, 1962), 89–92; F. W. Sears, *Thermodynamics* (Cambridge, MA: Addison-Wesley, 1953), 240–3.

13. Spector, "Models and Theories," in *Readings in Philosophy of Science*, 285.

14. Noretta Koertge, "For and Against Method," *Brit. J. Phil. Sci. 23* (1972), 275.

15. Carl Hempel, "On the 'Standard Conception' of Scientific Theories," in *Minnesota Studies in the Philosophy of Science, vol. IV*, ed. M. Radner and S. Winokur (Minneapolis: University of Minnesota Press, 1970), 142–63.

16. Ibid., 162.

17. Hempel, "Formulation and Formalization of Scientific Theories," in *The Structure of Scientific Theories*, ed. F. Suppe (Urbana: University of Illinois Press, 1974), 253.

18. Ibid., 245.

Doubts about the Chinese-Box View of Scientific Progress (pp. 144–155)

1. Paul K. Feyerabend, "Explanation, Reduction and Empiricism," in *Minnesota Studies in the Philosophy of Science, vol. III*, ed. H. Feigl and G. Maxwell (Minneapolis: University of Minnesota Press, 1963), 44.

2. Carl Hempel, *Aspects of Scientific Explanation* (New York: Free Press, 1965), 33.

3. Feyerabend, "Explanation, Reduction and Empiricism," 46–8.

4. Feyerabend, "On the 'Meaning' of Scientific Terms," *J. Phil. 62* (1965), 267–71; "Consolations for the Specialist," in *Criticism and the Growth of Knowledge*, ed. I. Lakatos and A. Musgrave (Cambridge: Cambridge University Press, 1970), 220–1; "Against Method: Outline of an Anarchistic Theory of Knowledge," in *Minnesota Studies in the Philosophy of Science, vol. IV*, ed. M. Radner and S. Winokur (Minneapolis: University of Minnesota Press, 1970), 14.

5. Feyerabend, "Explanation, Reduction, and Empiricism," 76–81.

6. Hilary Putnam, "How Not to Talk about Meaning," in *Boston Studies in the Philosophy of Science, vol. II*, ed. R. Cohen and M. Wartofsky (New York: Humanities Press, 1965), 206–7.

7. Feyerabend, "Reply to Criticism: Comments on Smart, Sellars and Putnam," in *Boston Studies in the Philosophy of Science, vol. II*, 229–30.

8. David Bohm, *Causality and Chance in Modern Physics* (New York: Harper Torchbooks, 1957), chapter IV; J.-P. Vigier, "The Concept of Probability in the Frame of the Probabilistic and Causal Interpretation of Quantum Mechanics," in *Observation and Interpretation*, ed. S. Korner (New York: Academic Press, 1957).

9. Feyerabend, "How to be a Good Empiricist—A Plea for Tolerance in Matters Epistemological," in *Readings in the Philosophy of Science*, ed. B. Brody (Englewood Cliffs, NJ: Prentice-Hall, 1970), 320–1.

10. Feyerabend, "Against Method," 26.

11. Feyerabend, "Problems of Empiricism, Part II," in *The Nature and Function of Scientific Theories*, ed. R. Colodny (Pittsburgh, PA: University of Pittsburgh Press, 1970), 307–23.

12. Galileo, *Dialogue Concerning the Two Chief World Systems*, trans. S. Drake (Berkeley: University of California Press, 1962), 173.

13. Feyerabend, "Explanation, Reduction, and Empiricism," 59.

14. Feyerabend, "Problems of Empiricism," in *Beyond the Edge of Certainty*, ed. R. Colodny (Englewood Cliffs, NJ: Prentice-Hall, 1965), 180.

15. Dudley Shapere, "Meaning and Scientific Change," in *Mind and Cosmos*, ed. R. Colodny (Pittsburgh, PA: University of Pittsburgh Press, 1966), 55–6.

16. Peter Achinstein, "On the Meaning of Scientific Terms," *J. Phil. 61* (1964), 504–5.

17. Ibid., 499.

18. Feyerabend, "Comments and Criticism on the 'Meaning' of Scientific Terms," *J. Phil. 62* (1965), 267.

19. Ibid.

20. Ibid., 268–69.

21. Shapere, "Meaning and Scientific Change," 64.

22. Feyerabend, "Problems of Empiricism," 216–17.

23. Feyerabend, "Against Method," 91.

24. Feyerabend, "Problems of Empiricism," 214.

25. Ibid., 198.

26. Ibid., 212.

27. Shapere, "Meaning and Scientific Change," 60.

28. Ibid.

29. Feyerabend, "Against Method," 91.

30. William Whewell, *History of the Inductive Sciences, vol. 1* (New York: D. Appleton, 1859), 47.

31. Stephen Toulmin, *Foresight and Understanding* (New York: Harper Torchbooks, 1961), 79.

32. N. R. Hanson, *Patterns of Discovery* (Cambridge: Cambridge University Press), chapter IV.

33. Ibid., 5–24.

34. Thomas S. Kuhn, *The Structure of Scientific Revolutions*, 1st edn. (Chicago: University of Chicago Press, 1962).

35. Kuhn, *The Structure of Scientific Revolutions*, 149.

36. Israel Scheffler, *Science and Subjectivity* (New York: Bobbs-Merrill, 1960), 19.

37. Gerd Buchdahl, "A Revolution in Historiography of Science," *Hist. Sci. 4* (1965), 55–69; Dudley Shapere, "The Structure of Scientific Revolutions," *Phil. Rev. 73* (1964), 383–94.

38. Kuhn "Postscript," in *The Structure of Scientific Revolutions*, 2nd edn. (Chicago: University of Chicago Press, 1970), 174–91.

39. Ibid., 175.

40. Ibid., 180–1.

41. Imre Lakatos, "Falsification and the Methodology of Scientific Research Programmes," in *Criticism and the Growth of Knowledge*, ed. I. Lakatos and A. Musgrave (Cambridge: Cambridge University Press, 1970); Larry Laudan, *Progress and Its Problems* (Berkeley: University of California Press, 1977); see also John Losee, *Theories of Scientific Progress* (London: Routledge, 2004).

The Legacy of Logical Reconstructionist Philosophy of Science (pp. 155–158)

1. Paul Feyerabend, "Philosophy of Science: A Subject with a Great Past," in *Historical and Philosophical Perspectives on Science*, ed. R. Stuewer (Minneapolis: University of Minnesota Press, 1970), 181.

2. Ibid., 183.

3. Ibid.

4. Feyerabend, *Against Method* (London: NLB, 1975), 23–4.

5. Herbert Feigl, "Empiricism At Bay?" in *Boston Studies in the Philosophy of Science, Vol. XIV*, ed. R. Cohen and M. Wartofsky (Dordrecht; Reidel, 1974), 8–9.

6. Ernest Nagel, *The Structure of Science* (New York: Harcourt, Brace & World, 1961), 86–8.

7. Gerald Holton, *The Advancement of Science* (Cambridge: Cambridge University Press, 1986), 163.

Chapter 3: Interlude: A Classificatory Matrix for Philosophies of Science (pp. 159–162)

1. William Whewell, *Philosophy of the Inductive Sciences*, 2nd edn., 1847 (London: Cass, 1967), vol. 2, 65.

2. John Stuart Mill, *A System of Logic*, 8th edn. (London: Longman, 1970), 323–4.

3. Carl Hempel, *Aspects of Scientific Explanation* (New York: Free Press, 1965), 13.

4. Imre Lakatos, "Falsification and the Methodology of Scientific Research Programmes," in *Criticism and the Growth of Knowledge*, ed. I. Lakatos and A. Musgrave (Cambridge: Cambridge University Press, 1970).

5. Larry Laudan, *Progress and Its Problems* (Berkeley: University of California Press, 1977), 158–64.

Chapter 4: Descriptivism (pp. 163–164)

1. Gerald Holton, "Do Scientists Need a Philosophy?" *Times Literary Supplement 2* (Nov. 1984), 1231–4.

Meehl's "Actuarial" Approach (pp. 164–165)

1. Paul Meehl, "The Miracle Argument for Realism: An Important Lesson to be Learned by Generalizing from Carrier's Counterexamples," *Stud. Hist. Phil, Sci. 23* (1992), 279.

2. Ibid., 281.

Holton on Thematic Principles (pp. 165–167)

1. Gerald Holton, "Thematic Presuppositions and the Direction of Scientific Advance," in *Scientific Explanation*, ed. A. F. Heath (Oxford: Clarendon Press, 1981), 17–23; *Thematic Origins of Scientific Thought*, revised edn. (Cambridge, MA: Harvard University Press, 1988); *The Scientific Imagination* (Cambridge: Cambridge University Press), 6–22.

2. Holton, *Thematic Origins of Scientific Thought*, 28–9; *The Scientific Imagination*, 6–22.

3. Holton, "Thematic Presuppositions," 7.

4. Ibid., 13–15.

Giere's "Satisficing Model" (pp. 167–168)

1. Ronald N. Giere, "Philosophy of Science Naturalized," *Phil. Sci. 52* (1985), 353.

Thagard's Computational Philosophy of Science (pp. 168–169)

1. Paul Thagard, "Computational Models in the Philosophy of Science," *PSA 1986*, ed. A. Fine and P. Machamer (East Lansing, MI: Philosophy of Science Association, 1987), 329–35; *Computational Philosophy of Science* (Cambridge, MA: MIT Press, 1988).

2. Ibid.
3. Herbert A. Simon, "Scientific Discovery as Problem Solving," *International Stud. Phil. Sci. 6* (1992), 5–12.
4. Dudley Shapere, "Discovery, Rationality, and Progress in Science: A Perspective in the Philosophy of Science," *PSA 1972*, in *Boston Studies in the Philosophy of Science, vol. 20*, ed. K. Schaffner and R. S. Cohen (Dordrecht: Reidel, 1974), 409–19; "Scientific Theories and Their Domains," in *The Structure of Scientific Theories*, ed. F. Suppe (Urbana: University of Illinois Press), 549–55.

Cognitive Development and Theory-Change (pp. 169–172)

1. Alison Gopnik, "The Scientist as Child," *Phil. Sci. 63* (1996), 485–514.
2. Ibid., 486.
3. Ibid., 485.
4. Ibid., 487.
5. Ronald N. Giere, "The Scientist as Adult," *Phil. Sci. 63*, (1996), 538–41.
6. Gopnik, "Reply to Commentators," *Phil. Sci. 63* (1996), 554.
7. Susan Carey and Elizabeth Spelke, "Science and Core Knowledge," *Phil. Sci. 63* (1996), 517–20.

The Evolutionary-Analogy View (pp. 172–196)

1. Michael Ruse, *Taking Darwin Seriously* (Oxford: Blackwell, 1986), 29–66, 149–68.
2. Stephen Toulmin, *Human Understanding* (Oxford: Clarendon Press, 1972), 136.
3. Toulmin, *Foresight and Understanding* (New York: Harper Torchbooks, 1961), 110.
4. Toulmin, *Human Understanding, passim.*
5. Toulmin, "Rationality and Scientific Discovery," in *Boston Studies in the Philosophy of Science, vol. XX*, ed. R. Schaffer and R. Cohen (Dordrecht: Reidel, 1974), 394.
6. Ibid., 391.
7. Ernan McMullin, "Logicality and Rationality, a Comment on Toulmin's Theory of Science," in *Philosophical Foundations of Science*, ed. R. J. Seeger and R. S. Cohen (Dordrecht: Reidel, 1974), 427.
8. Toulmin, "Rationality and Scientific Discovery," 395.
9. Ibid., 400.
10. Toulmin, *Human Understanding*, 139.
11. Ibid., 141.
12. David Hull, *Science as a Process* (Chicago: University of Chicago Press, 1988), 409; *The Metaphysics of Evolution* (Albany, NY: SUNY Press, 1989), 96.
13. Hull, *The Metaphysics of Evolution*, 221.
14. Ibid., 105.

15. Ibid., 221.
16. Ibid., 234.
17. Patrick Matthew, *On Naval Timbre and Aboriculture* (London: Longman, 1831).
18. Hull, *The Metaphysics of Evolution*, 235.
19. Ibid., 124.
20. Ibid., 235–7.
21. Moritz Wagner, *The Darwinian Theory and the Law of the Migration of Organisms* (London: Stanford, 1872).
22. Ernst Mayr, *Systematics and the Origin of Species* (New York: Columbia University Press, 1942).
23. Richard Goldschmidt, *The Material Basis of Evolution* (New Haven, CT: Yale University Press, 1940).
24. Nels Eldredge and Stephen J. Gould, "Punctuated Equilibria," in *Models in Paleobiology*, ed. T. J. M. Schopf (San Francisco: Freeman, Copper & Co., 1972).
25. Hull, *Metaphysics of Evolution*, 236.
26. J. W. van Spronsen, *The Periodic System of the Chemical Elements* (Amsterdam: Elsevier, 1969), 147–209.
27. Richard Dawkins, *The Selfish Gene*, 1976, 30th Anniversary Edition (Oxford: Oxford University Press, 2006), 254.
28. Ibid.
29. Ibid., 28.
30. Stephen Jay Gould, *The Panda's Thumb* (New York: W. W. Norton, 1980), 90.
31. Dawkins, *The Selfish Gene*, 235.
32. Dawkins, *The Selfish Gene*; *The Blind Watchmaker* (New York: W. W. Norton & Co., 1987); *Climbing Mount Improbable* (New York: W. W. Norton & Co., 1997); *The Extended Phenotype* (Oxford: Oxford University Press, 1999); *The Greatest Show on Earth* (New York: Free Press, 2010).
33. Dawkins, *The Selfish Gene*, 254.
34. Ibid., 9–11.
35. Ibid., 88.
36. Ibid., 243.
37. Ibid., 59.
38. Ibid., 60.
39. Ibid., 189.
40. Ibid., 192.
41. Ibid.
42. Ibid., 194.
43. Ibid.
44. Ibid., 196.
45. Ibid., 197.

Notes

46. Dawkins, *The Extended Phenotype* (Oxford: Oxford University Press, 1999), 109.
47. Ibid.
48. Ibid., 112.
49. Ibid.
50. Ibid.
51. Ibid.
52. Ibid.
53. Ibid., 4.
54. Ibid., 112–17.
55. Ibid., 112.
56. Joseph Fracchia and Richard C. Lewontin, "Does Culture Evolve?" in *Conceptual Issues in Evolutionary Biology*, ed. E. Sober (Cambridge, MA: MIT Press, 2006), 505–33.
57. Dawkins, *The Extended Phenotype*, 112.
58. Fracchia and Lewontin, "Does Culture Evolve?" 524.
59. Ibid., 525.
60. Ibid., 505.
61. L. Jonathan Cohen, "Is the Progress of Science Evolutionary?" *Brit. J. Phil. Sci. 24* (1973), 47.
62. Donald T. Campbell, "Blind Variation and Selective Retention in Creative Thought as in Other Knowledge Processes," *Psych. Rev. 67* (1960), 380; reprinted in *Evolutionary Epistemology, Rationality, and the Sociology of Knowledge*, ed. G. Radnitzky and W. W. Bartley, III (La Salle, IL: Open Court, 1987), 91.
63. Ron Amundson, "The Trials and Tribulations of Selectionist Explanations," in *Issues in Evolutionary Epistemology*, ed. K. Hahlweg and C. A. Hooker (Albany, NY: SUNY Press, 1989), 428.
64. Campbell, "Blind Variation and Selective Retention," 91.
65. Karl Popper, "Replies to my Critics," in *The Philosophy of Karl Popper*, ed. P. A. Schilpp (La Salle, IL: Open Court, 1974), 1061.
66. Campbell, "Blind Variation and Selective Retention," 92–3.
67. Ruse, *Taking Darwin Seriously*, 65.

The Evolutionary-Origins View (pp. 196–205)

1. Michael Ruse, *Evolutionary Naturalism* (London: Routledge, 1995), 159–60.
2. D. M. Dallas, "The Chemical Calculus of Sir Benjamin Brodie," in *The Atomic Debates* (Leicester: Leicester University Press, 1967), 37.
3. Gerald Holton, *Thematic Origins of Scientific Thought* (Cambridge, MA: Harvard University Press, 1988), 253.
4. Ruse, *Evolutionary Naturalism*, 169.
5. Ibid., 164.

The Strong Programme (pp. 205–208)

1. David Bloor, *Knowledge and Social Imagery*, 2nd edn. (Chicago: University of Chicago Press, 1991), 7.
2. Barry Barnes, *Scientific Knowledge and Sociological Theory* (London: Routledge & Kegan Paul, 1974), 120.
3. Steven Shapin and Simon Schaffer, *Leviathan and the Air-Pump* (Princeton, NJ: Princeton University Press, 1985).
4. Martin Rudwick, *The Great Devonian Controversy* (Chicago: University of Chicago Press, 1985).
5. Barry Barnes, David Bloor, and John Henry, *Scientific Knowledge* (Chicago: University of Chicago Press, 1996), 18–45.
6. Ibid., 25.
7. Michael Friedman, "On the Sociology of Scientific Knowledge and Its Philosophical Agenda," *Stud Hist. Phil. Sci. 29A* (1998), 239–71.
8. Bloor, *Knowledge and Social Imagery*, 5.
9. Mario Biagoli, *Galileo, Courtier* (Chicago: University of Chicago Press, 1993); Shapin and Schaffer, *Leviathan and the Air-Pump*; Martin Rudwick, *The Great Devonian Controversy* (Chicago: University of Chicago Press, 1985); Andrew Pickering, *Constructing Quarks* (Edinburgh: University of Edinburgh Press, 1984).

Chapter 5: Normative Naturalism

Neurath's "Boat-Repair" Image (pp. 209–212)

1. Otto Neurath, "Protocol Statement," in *Otto Neurath: Philosophical Papers*, ed. R. S. Cohen and M. Neurath (Dordrecht: Reidel, 1983), 92.
2. Nancy Cartwright, Jordi Cat, Lola Fleck, and Thomas E. Uebel, in *Otto Neurath: Philosophy between Science and Politics* (Cambridge: Cambridge University Press, 1996), 130–48, identify three separate "Boat Images" in Neurath's writings.
3. See, for instance, John Losee, *Complementarity, Causality and Explanation* (New Brunswick, NJ: Transaction Publishers, 2013), 53–102.

Quine's "Field-of-Force Image" (p. 212)

1. Willard van Orman Quine, "Two Dogmas of Empiricism," in *From a Logical Point of View* (Cambridge, MA: Harvard University Press, 1953), 79.
2. Ibid., 44.

Kantorovich's Evolutionist Normative Naturalism (p. 213)

1. Aharon Kantorovich, *Scientific Discovery* (Albany, NY: SUNY Press, 1993), 117.
2. J. J. C. Smart, "Science, History and Methodology," *Brit. J. Phil. Sci. 23* (1972), 268.
3. Kantorovich, *Scientific Discovery*, 131.
4. Ibid., 125.
5. Ibid., 126.

Local and Global Applications of Evaluative Standards (pp. 214–218)

1. Arthur Fine, "The Natural Ontological Attitude," in *Scientific Realism*, ed. J. Leplin (Berkeley: University of California Press, 1984), 96.
2. Richard M. Burian, Robert C. Richardson, and Wim J. van der Steen, "Against Generality: Meaning in Genetics and Philosophy," *Stud. Hist. Phil. Sci. 27* (1996), 3.
3. Ibid., 12.
4. Ibid., 8.
5. Ibid.
6. Ibid., 9.
7. Ibid., 19.
8. Ibid., 25.
9. Ibid.
10. Ibid., 26.
11. Ibid.

Philip Kitcher on Achieving Cognitive Virtue (pp. 218–220)

1. Philip Kitcher, "The Naturalists Return," *Phil. Rev. 101* (1992), 77.
2. Ibid., 97–8.
3. P. W. Bridgman, *The Logic of Modern Physics* (New York: Macmillan, 1927), 3–14; *The Nature of Physical Theory* (New York: Dover, 1936), 7–9.
4. Kitcher, "Explanatory Unification and the Causal Structure of the World," in *Scientific Explanation* ed. P. Kitcher and W. Salmon (Minneapolis: University of Minnesota Press, 1989), 430–7.
5. Kitcher, *The Advancement of Science* (New York: Oxford University Press, 1993) 157–60.

Shapere's Nonpresuppositionist Philosophy of Science (pp. 220–228)

1. Dudley Shapere, "The Character of Scientific Change," in *Scientific Discovery, Logic, and Rationality,* ed. T. Nickles (Dordrecht: Reidel, 1980), 94.

2. Shapere, "Scientific Theories and Their Domains," in *The Structure of Scientific Theories*, ed. F. Suppe (Urbana, IL: University of Illinois Press, 1974), 525.

3. Shapere, "Discovery, Rationality and Progress in Science: A Perspective in the Philosophy of Science," in *Boston Studies in the Philosophy of Science, vol. XX*, ed. K. Schaffner and R. S. Cohen (Dordrecht: Reidel, 1974), 409.

4. Shapere, "Scientific Theories and Their Domains," 549–55.

5. I. Bernard Cohen, "Discussion: Shapere on Scientific Theories and Their Domains," in *The Structure of Scientific Theories*, 590–3.

6. Shapere, "Discovery, Rationality and Progress in Science," 419.

7. Ibid., 411–12.

8. Shapere, *Reason and the Search for Knowledge* (Dordrecht: Reidel, 1983), 10.

9. Shapere, "The Character of Scientific Change," 72–3; "Reply to 'Discussion of Shapere's Paper: "The Character of Scientific Change,"'" in *Scientific Discovery, Logic, and Rationality*, 103.

10. Shapere, *Reason and the Search for Knowledge*, 12.

11. Shapere, "The Character of Scientific Change," 83.

12. Ibid., 83–4.

13. Ibid., 8.

14. Paul Feyerabend, *Against Method* (London: NLB, 1974), 23.

15. Shapere, "The Character of Scientific Change," 63.

16. Shapere, *Reason and the Search for Knowledge*, 27.

17. Ibid., 23.

18. Ibid., 23, 31.

19. Ibid., 23.

20. Ibid., 17.

21. Ibid., 36–41.

22. Shapere, "Reason, Reference and the Quest for Knowledge," *Phil. Sci. 49* (March 1982), 21.

23. Shapere, "The Concept of Observation in Science and Philosophy" *Phil. Sci. 49* (December 1982), 31–2.

24. Shapere, "Reply," 108.

25. Larry Laudan, "Discussion of Shapere's Paper: 'The Character of Scientific Change,'" 103–4.

26. Shapere, "Reply," 106.

27. Shapere, "The Character of Scientific Change," 68.

28. Shapere, "Reply," 102.

29. Ibid.

30. Imre Lakatos, "Falsification and the Methodology of Scientific Research Programmes," in *Criticism and the Growth of Knowledge*, ed. I. Lakatos and A, Musgrave (Cambridge: Cambridge University Press, 1970), 116–22, 154–9; "History

of Science and Its Rational Reconstructions," in *Boston Studies in the Philosophy of Science, vol. VIII*, ed. R. Buck and R. S. Cohen (Dordrecht: Reidel, 1971), 93–136.

31. Larry Laudan, *Progress and Its Problems* (Berkeley: University of California Press, 1977), 155–70.
32. Shapere, *Reason and the Search for Knowledge*, 42.

Laudan's Three Versions of Normative Naturalism (pp. 228–244)

1. Larry Laudan, *Progress and Its Problems* (Berkeley: University of California Press, 1977), 59.
2. Ibid., 15.
3. Ibid., 17.
4. Ibid., 29.
5. Ibid., 63.
6. Ibid., 48.
7. Ibid., 55.
8. Isaac Newton, *Mathematical Principles of Natural Philosophy*, trans. A. Motte, revised F. Cajori (Berkeley: University of California Press, 1962), vol. II, 547.
9. Laudan, *Progress and Its Problems*, 68.
10. Ibid., 130.
11. Ibid., 161.
12. Ibid., 163.
13. Laudan, *Progress and Its Problems*, 161.
14. Ibid., 162–3.
15. Ibid., 130.
16. Ibid., 81.
17. Paul Feyerabend, "Consolations for the Specialist," in *Criticism and the Growth of Knowledge*, ed. I. Lakatos and A. Musgrave (Cambridge: Cambridge University Press, 1970), 215.
18. Imre Lakatos, "History of Science and Its Rational Reconstruction," in *Boston Studies in the Philosophy of Science, vol. VIII*, ed. R. Buck and R. S. Cohen (Dordrecht: Reidel, 1971), 104.
19. Laudan, *Progress and Its Problems*, 106–14.
20. Feyerabend, *Against Method* (London: NLB, 1975), 23.
21. Laudan, *Science and Values* (Berkeley: University of California Press, 1984).
22. Ibid., 57–9.
23. Ibid., 59.
24. William Gilbert, *De Magnete* (New York: Dover, 1958).
25. Newton, *Mathematical Principles of Natural Philosophy*, vol. II, 398–400; 547.

26. Gerald Holton, *The Advancement of Science, and Its Burdens* (Cambridge: Cambridge University Press, 1986), 68–76; *Thematic Origins of Scientific Thought*, revised edn. (Cambridge, MA: Harvard University Press, 1988), 252–4.

27. Gerald Doppelt, "Relativism and the Reticulational Model of Scientific Rationality," *Synthese 69* (1986), 234–7.

28. Laudan, *Science and Values*, 104.

29. Paul Thagard, "From the Descriptive to the Normative in Psychology and Logic," *Phil. Sci. 49* (1982), 25.

30. Laudan, "Progress or Rationality? The Prospects for a Normative Naturalism," *Amer. Phil. Quart. 24* (1987), 24.

31. Ibid., 23.

32. Ibid., 25.

33. Robert Nola and Howard Sankey, *Theories of Scientific Method* (London: Routledge, 2014), 324.

34. Laudan, "Progress or Rationality?" 26.

35. Ibid., 19.

Chapter 6: Foundationalism

General Arguments against Foundationalism (pp. 245–252)

1. Robert Audi, *The Structure of Justification* (Cambridge: Cambridge University Press, 1993), 3.

2. Bas van Fraassen, "Sola Experientia? Feyerabend's Refutation of Classical Empiricism," *PSA 1996 Part 2, Phil. Sci. 64, no. 4, Supplement* (1997), S385–S395.

3. Ibid., S388.

4. Ibid., S391.

5. See, for instance, John Losee, *Theories of Scientific Progress* (New York: Routledge, 2004).

6. Gerald Doppelt, "The Naturalist Conception of Methodological Standards in Science," *Phil. Sci. 57* (1990), 1.

7. Ibid., 11.

8. Ibid., 11–12.

9. Larry Laudan, *Science and Values* (Berkeley: University of California Press, 1984), 35–36, 84.

The "Strong Foundationalist" Position (p. 252)

1. John Worrall, "Fix It and Be Damned: A Reply to Laudan," *Brit. J. Phil. Sci. 40* (1989), 377.

Candidates for Foundational Status (pp. 253–260)

1. Imre Lakatos, "History of Science and Its Rational Reconstructions," *Boston Studies in the Philosophy of Science, vol. VIII*, ed. R. C. Buck and R. S. Cohen (Dordrecht: Reidel, 1971), 113.
2. P. K. Feyerabend, "An Attempt at a Realistic Interpretation of Experience," *Proc. Arist. Soc. 58* (1958), 160–2; Peter Achinstein, *Concepts of Science* (Baltimore, MD: The Johns Hopkins Press, 1968), 160–72; W. van Orman Quine, *From A Logical Point of View* (Cambridge, MA: Harvard University Press, 1953), 41.
3. Bas Van Fraassen, *The Scientific Image* (Oxford: Clarendon Press, 1980), 46–7.
4. Moritz Schlick, "The Foundation of Knowledge," in *Logical Positivism*, ed. A. J. Ayer (Glencoe, IL: Free Press, 1959) 209–27.
5. See Thomas Oberdan, "The Vienna Circle's Anti-Foundationalism," *Brit. J. Phil. Sci. 49* (1998), 302–6.
6. Karl Popper, *The Logic of Scientific Discovery* (New York: Basic Books, 1959), 43–8.
7. Ibid., 109–11.
8. Feyerabend, "An Attempt at a Realistic Interpretation,"; Achinstein, *Concepts of Science*; Quine, *From A Logical Point of View*.
9. Popper, *The Logic of Scientific Discovery*, 81–4.
10. Dmitri Mendeleeff, "The Periodical Law of the Chemical Elements," *Chemical News XL, no. 1045* (1879); reprinted in *Classical Scientific Papers—Chemistry, Second Series*, ed. D. M. Knight (London: *Mills and Boon*, 1970), 268.
11. Gerald Holton, "Thematic Presuppositions and the Direction of Scientific Advance," in *Scientific Explanation*, ed. A. F. Heath (Oxford: Clarendon Press, 1981), 15.
12. Morton Grosser, *The Discovery of Neptune* (Cambridge, MA: Harvard University Press, 1962), 58–98.
13. Isaac Asimov, *The Neutrino* (Garden City, NY: Doubleday, 1966), 113–42; Max Born, *Atomic Physics*, 5th edn. (New York: Hafner, 1950), 66–7.
14. Alan Musgrave, "Method or Madness?" *Boston Studies in the Philosophy of Science, vol. 39*, ed. R. S. Cohen et al. (Dordrecht: Reidel, 1976), 460.
15. Fred Hoyle, *Astronomy* (London: Crescent Books, 1962), 169.
16. David Miller, *Critical Rationalism* (Chicago: Open Court, 1994), 48.
17. Ibid., 48–9.
18. Nelson Goodman, *Fact, Fiction and Forecast*, 2nd edn. (Indianapolis, IN: Bobbs-Merrill, 1965), 74.

Empirical Laws as Foundational (pp. 260–265)

1. Ernest Nagel, *The Structure of Science* (New York: Harcourt, Brace & World, 1961), 86.

2. Herbert Feigl, "Empiricism At Bay?" *Boston Studies in the Philosophy of Science, vol. XIV*, ed. R. S. Cohen and M. Wartofsky (Dordrecht: Reidel, 1974), 9–10.

3. Wilfrid Sellars, "The Language of Theories," in *Current Issues in the Philosophy of Science* (New York: Holt, Rinehart and Winston, 1961), 71–2.

4. Ronald Giere, "Testing Theoretical Hypotheses," in *Minnesota Studies in the Philosophy of Science, vol. X*, ed. J. Earman (Minneapolis: University of Minnesota Press, 1983), 269–98.

5. Nancy Cartwright, *How the Laws of Physics Lie* (Oxford: Clarendon Press, 1983), 3.

6. Ibid., 57.

7. Ibid., 54.

8. Ian Hacking, *Representing and Intervening* (Cambridge: Cambridge University Press, 1983), 274.

9. Rom Harre, "Three Varieties of Realism," in *The Scientific Realism of Rom Harre*, ed. A. Derksen (Tilburg: Tilburg University Press, 1994), 15.

10. Harre, *Varieties of Realism* (Oxford: Blackwell, 1986).

11. Robert Brandon, *Concepts and Methods in Evolutionary Biology* (Cambridge: Cambridge University Press, 1996), 28.

12. Brandon, *Concepts and Methods*, 23: *Adaptation and Environment* (Princeton, NJ: Princeton University Press, 1990), 15.

13. Brandon, *Adaptation and Environment*, 158.

Evaluative Standards as Foundational (pp. 265–271)

1. Thomas S. Kuhn, *The Essential Tension* (Chicago: University of Chicago Press, 1977), 322–3.

2. Ibid., 331.

3. Norman R. Campbell, *Foundations of Science* (New York: Dover 1957), 123–4.

4. William Whewell, *Philosophy of the Inductive Sciences, Part II, Book X* (London: Cass, 1967), book II, 65.

5. Ibid., 67–8.

6. Jean Perrin, *Atoms*, trans. D. L. Hammick (London: Constable, 1922), 214–16.

7. Max Planck, *A Survey of Physics* (London: Methuen, 1925), 163–76.

8. Imre Lakatos, "Changes in the Problem of Inductive Logic," in *Inductive Logic*, ed. I. Lakatos (Amsterdam: North-Holland, 1968); "Falsification and the Methodology of Research Programmes" in *Criticism and the Growth of Knowledge*, ed. I. Lakatos and A. Musgrave (Cambridge: Cambridge University Press, 1970); "History of Science and Its Rational Reconstructions," in *Boston Studies in the Philosophy of Science, vol. VIII*, ed. R. Buck and R. S. Cohen (Dordrecht: Reidel, 1971).

9. Lakatos, "Falsification and the Methodology of Scientific Research Programmes," 116–18.

10. Lakatos, "History of Science and Its Rational Reconstructions." 132–8.
11. P. K. Feyerabend, *Against Method*, (London: NLB, 1975), 175–80.
12. See, for instance, C. Prutton and S. Maron, *Fundamental Principles of Physical Chemistry* (NewYork: Macmillan, 1951), 210–24, 460–70.
13. Lakatos, "Falsification and the Methodology of Scientific Research Programmes," 138–40.
14. J. B. A. Dumas, untitled, *Ann. Chim. Phys. 55* (1859), 129.

Cognitive Aims as Foundational (pp. 271–276)

1. Ernan McMullin, "Rationality and Paradigm Change in Science," in *World Changes*, ed. P. Horwich (Cambridge: MIT Press, 1993), 67–70.
2. Ibid., 69.
3. Ibid.
4. Andreas Osiander, Preface to Copernicus, *De revolutionibus*, in *Three Copernican Treatises*, ed. E. Rosen (New York: Dover, 1959), 24–5.

Methodological Rules as Foundational (pp. 276–277)

1. John Worrall, "The Value of a Fixed Methodology," *Brit. J. Phil. Sci. 39* (1988), 274.
2. Worrall, "Fix It and Be Damned: A Reply to Laudan," *Brit. J. Phil. Sci. 40* (1989), 386.
3. Ibid., 380.
4. Lewis Carroll, "What the Tortoise Said to Achilles," *Mind 13* (1895), 278–80.
5. Larry Laudan, "If It Ain't Broke, Don't Fix It," *Brit. J. Phil. Sci. 40* (1989), 375.

Bayesian Conditionalization (pp. 277–80)

1. Thomas Bayes, "An Essay towards Solving a Problem in the Doctrine of Chances," *Phil. Trans. 53* (1763), 370–418, reprinted in *Biometrika 45* (1958), 296–315.
2. Colin Howson and Peter Urbach, *Scientific Reasoning: The Bayesian Approach* (La Salle, IL: Open Court, 1989), 270–5.
3. Daniel Garber, "Old Evidence and Logical Omniscience in Bayesian Confirmation Theory," in *Testing Scientific Theories*, ed. J. Earman (Minneapolis: University of Minnesota Press, 1983), 99–131.
4. Richard W. Miller, *Fact and Method* (Princeton, NJ: Princeton University Press, 1987), 297–319.

Chapter 7: Descriptivism Versus Foundationalism

The Case for Descriptivism (pp. 281–283)

1. Friedrich Waismann, "How I See Philosophy," in Logical Positivism, ed. A. J. Ayer (Glencoe: Free Press, 1959), 345.

The Case for Foundationalism (pp. 283–284)

1. John Worrall, "Fix It and Be Damned: A Reply to Laudan," *Brit. J. Phil. Sci. 40* (1989), 383.

Index of Names

Index of Subjects